Fiber Bragg Grating Sensors

Development and Applications

Fiber Bragg Grating Sensors

Development and Applications

Authored by
Hisham K. Hisham

CRC Press
Taylor & Francis Group
Boca Raton London New York

CRC Press is an imprint of the
Taylor & Francis Group, an **informa** business

Taylor & Francis Group
6000 Broken Sound Parkway NW, Suite 300
Boca Raton, FL 33487-2742

First issued in paperback 2023

ISBN-13: 978-0-367-22485-1 (hbk)
ISBN-13: 978-1-03-265401-0 (pbk)
ISBN-13: 978-0-429-27513-5 (ebk)

DOI: 10.1201/9780429275135

Dedication

To my parents,
To my beloved wife and my children.

Contents

Preface

Sensing technologies based on optical fiber have many inherent advantages that make it an important and attractive option for a wide range of industrial sensing applications. Many have been published in recent years, including a number of papers and books. However, given the importance of this advanced and growing technology in various fields, it is clear that there was need for a more comprehensive book of various industrial applications. The tremendous development of this technology in various fields made the timing of this book extremely important, raising the challenge. However, the efforts made to collect many research and studies in this field have made us believe that this is applicable and enabled us to create a great multi-faceted idea and present it fully. Therefore, our goal was to provide a comprehensive and up-to-date overview of the subject, which is the basis for building future works. As part of this ideal we have included more than 400 references. This work is primarily suitable for the researcher or academic in the field of optical sensing technology; however, its independent form is generally suited to engineers or graduate students and requires only basic knowledge of light properties and fiber Bragg gratings (FBGs) structure.

The book begins with a brief introduction to fiber Bragg gratings, followed by the detailed explanation of physical description, photosensitivity, types, and manufacturing techniques.

Chapter 2 provides a detailed explanation to the polymer optical fiber (POF). A brief comparison between silica optical fiber (SOF) and polymer optical fiber Bragg gratings is made. The polymer properties, photosensitivity, and grating's manufacturing are dealt with Chapter 2. The tuning characteristics of fiber grating and a brief summary for theory that describes the propagation characteristics of light waves in optical fibers are described in Chapter 2. The relatively simple fiber modes, optical parameters, coupled-mode theory, and modeling of fiber Bragg grating is also described in Chapter 2.

Chapter 3 is interested in discussing the basic details of Bragg grating properties (i.e. reflectivity, bandwidth, delay time, and dispersion) of both silica and polymer fibers. It also provides essential details regarding the temperature-induced wavelength shift properties. Chapter 4 informs the reader of the most important sensor mechanisms in fiber Bragg gratings, gratings simulation methods and developments, while also providing essential details for the treatment of temperature dependence of the strain sensitivity theory.

Chapter 5 details the basics of fiber optic sensors in the field of oil and gas applications; the essentials of optical fibers for downhole power transmission and data measuring remotely; and the distributed sensor for temperature (DTS), the acoustics (DAS) and the strain (DSS) measurements; the downhole monitoring (i.e. reservoir pressure and temperature monitoring, flow monitoring, seismic monitoring, pipeline monitoring, power cable and transformer monitoring, status monitoring of water mains). It also provides essential details regarding the fiber optic sensors for detection applications (i.e. leak detection, ground movement detection, fire detection, pig position detection) and the principles of gas network monitoring.

Chapter 6 explores fiber optic sensors in civil applications, an area where Bragg gratings have had a tremendous impact. The applications include those related to infrastructure sensors (i.e. crack, strain, corrosion). Also, this chapter provides information about the optical sensor application to highway structures (bridges, dams, buildings), geotechnical structures, historical buildings, pavements, tunnels, embankments, slopes, and other applications.

An increasingly important field is fiber optic sensors, and an expansive treatment of Bragg grating for biomedical applications is provided in Chapter 7. Essential information about the basic biomedical instrumentation system and the application fields of biomedical sensors (i.e. glucose sensor, laminate cure analysis, protein analysis, dosage form analysis, drug identification, determination of DNA oligomers, pesticide detection, effluent monitoring) and others are provided.

Chapter 8 discusses the use of optical fiber in one of the most important and sensitive areas of the sensor, the field of military applications. The chapter provides information about the benefits of the use of fiber optic technology in military communications systems and sensor fields in various military applications (i.e. communications, weapon systems, surveillance, aboard vehicles) and others.

Chapter 9 discusses the applications of fiber optic sensors in the field of harsh environments, the types of optical fibers that are suitable for harsh conditions, how to use it in the field of sensing, and how the measured data are transmitted. It also provides very important information about the most important harsh environmental sensing parameters (i.e. temperature, radiation, pressure, hydrogen detection, high strain measurements) and others.

Chapter 10 highlights the use of fiber optic sensors in other important industrial applications, for example, a ship's cargo handling system, longitudinal/transversal ship's hull strength monitoring, measuring the mass of cargo loaded or discharged, liquid level readings in different tanks, monitoring temperature/humidity in a ship's cargo, load forces and strain effects, and fiber optic sensing for satellite communications.

Acknowledgment

We are grateful to the following for permission to produce copyright materials:

Figures 1.1, 6.1, 6.2, 6.3, 6.4, 6.5, 6.6, 6.7, 6.8, and 6.9 from Applications of fiber optic sensors in civil engineering, *Struct. Eng. Mech.* 25(5), 577–596, 2007.

In some instances, we have been unable to trace the owners of copyright material, and we would appreciate any information that would enable us to do so.

I would like to express my great gratitude for the many comments and to the brilliant suggestions which were made by the dedicated reviewers who encouraged and stimulated improvements to the text.

I would also like to thank the authors of the many journals, conference papers, articles, theses, and books that have been relied on, as well as the references, publishers, and scientific institutions that have granted us permission to use drawings and pictures.

I am also very grateful to my family and to my friends who have shown their continued support and great interest over a long period of work on the book, and here I must mention my dear friend Dr. Ghafour Amouzad Mahdiraji.

Finally, I would like to thank Marc Gutierrez, for his patience, constant support, and dedication, which enabled me to complete this mission, and to all those who helped me to make this a success.

Hisham K. Hisham

Author

Hisham K. Hisham obtained his Ph.D degree from the Centre of Excellence for Wireless and Photonic Networks (CEWPN), Universiti Putra Malaysia (UPM), Malaysia in 2012. His Ph.D. work on fiber laser design produced a patent.

He has authored and coauthored several technical papers, which include journal articles and conference proceedings. His research interests are in optical communications, fiber lasers, devices, and sensors.

1 General Introduction

1.1 FIBER BRAGG GRATINGS

The discovery of photosensitivity in optical fibers [1] has had major role in the development of telecommunication and sensor systems technologies, the effect being used to develop devices for many applications. The construction of a fiber Bragg grating (FBG) is usually based on the photosensitivity property of silica fiber doped with germanium. Photosensitivity means that exposure of ultraviolet (UV) light results in a rise in the refractive index of certain doped glasses. Therefore, when exposed to a UV radiation, the fiber exhibits a permanent change in the refractive index of the core, which depends on the pattern and the properties of the UV exposure beam [2, 3].

When a side of fiber is exposed to a UV radiation by the interference of two intersecting beams, a FBG is fabricated with a user-defined central wavelength independent of the wavelength of the writing beam. This UV exposure of the fiber will write a regular form of a periodicity half the required Bragg wavelength into the fiber core over lengths in the range of millimeters to centimeters. This flexibility in the fabrication of FBGs allows Bragg wavelengths to change from the visible region to well beyond the telecommunications wavelength of 1550 nm to be written [2].

When the optical fiber refractive index is changed, the light that is transmitted through the fiber will be reflected back to the source; thereby, based on this property, FBGs serve as a wavelength-selective device within the optical fiber. Because of this selectivity, a narrow band of the incident field within the optical fiber is reflected by successive, coherent scattering from the index variations, transmitting the rest. This operation is shown in Figure 1.1. Due to fiber refractive index sensitivity, therefore, any change in the induced grating period will lead to a shift in the reflected and transmitted wavelength spectra. This property makes FBGs useful as tuning mechanisms [2, 3].

In recent years, FBGs have been a growing interest due to their ability in design for wide wavelength-selective range; therefore, they can be used in a variety of applications. For telecommunications, probably the most promising applications have been dispersion compensation and wavelength-selective devices such as filters for wavelength-division multiplexing (WDM) [4–8].

FBGs have also become popular as sensing devices, ranging from structural monitoring to chemical sensing [9–12]. Any change in the fiber properties, such as strain or temperature, changes the fiber index or grating pitch, which leads to changing the Bragg wavelength [13–18]. Therefore, by determining the peak reflectivity wavelength of the grating, we will get information about the sensing parameters. Another important application of fiber gratings is to use them as wavelength reflectors for fiber lasers [15, 18–20].

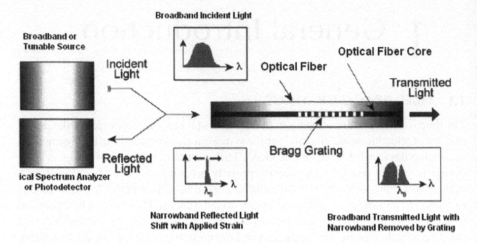

FIGURE 1.1   Fiber Bragg gratings as a wavelength-selective element within the optical fiber [37].

1.2   FUNDAMENTALS OF BRAGG GRATINGS

1.2.1   PHYSICAL DESCRIPTIONS

A Bragg grating formed by UV exposure without physical trenches is referred to as a bulk index grating, and a Bragg grating formed by physical deformation is referred to as a surface relief grating. Bragg gratings can be found in widely photonic components and systems, including distributed feedback laser diodes (DFB-LDs), distributed Bragg reflector laser diodes (DBR-LDs) and can be formed in optical fibers, as well as in planar waveguides [21]. Two of the most popular forms of Bragg gratings are shows in Figure 1.2 and 1.3; a fiber Bragg grating, i.e. a Bragg grating formed in a fiber, and a planar Bragg grating (PBG), i.e. a Bragg grating in a planar waveguide. An FBG is usually a bulk index grating, while both surface relief and bulk indexing are used to create a polymer Bragg grating [22–25].

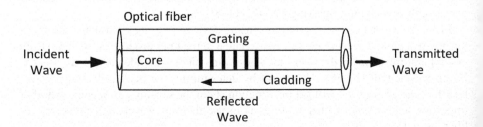

FIGURE 1.2   Fiber Bragg grating structure [37].

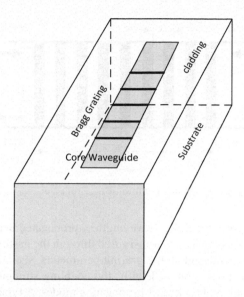

FIGURE 1.3   Planar Bragg grating structure.

1.2.2 Photosensitivity

Photosensitivity in optical fiber usually refers to the permanent change in the refractive index of the optical fiber core that is induced by exposure to highly UV light radiation [2]. Photosensitivity is a nonlinear effect, which can be seen as a growth in intensity of the light that is reflected from an optical fiber after the fiber was exposed to intense laser radiation. The discovery of this effect led to a revolutionary spread of the fiber Bragg grating structure. Widespread studies of this effect on an optical fiber have allowed an absolute control of the characteristics of the refractive index gratings.

There are several techniques for improving the optical fiber photosensitivity; some of these include: increased concentration of germanium doping, heat treating, boron co-doping, hydrogen loading, thermal-induced refractive index change, doping Ge Si fiber with boron and tin and hot hydrogenation, cold hydrogenation under high pressure [2].

1.3 TYPES OF FIBER BRAGG GRATINGS

There are several different kinds of FBG structures: the common Bragg grating reflector, the chirped Bragg grating, and the tilted Bragg grating [2, 3]. These FBGs are characterized either through the grating pitch or tilt.

1.3.1 Common Bragg Reflector

For a common Bragg reflector, the grating period (Λ) remains constant throughout the grating length and the reflection is the strongest at the Bragg wavelength λ_B. The Bragg resonant wavelength is a function of Λ and the mode effective reflection index

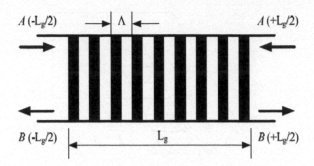

FIGURE 1.4   Schematic diagram of a common Bragg reflector.

(n_{eff}) [2]. Thus, the light at the Bragg wavelengthλ_B propagated in the fiber undergoes reflection and the rest of the light is transmitted through the grating unimpeded. The spectral characteristics depend on the grating parameters, such as the amplitude of the refractive modulations, grating length, the coupling strength, and the overlap integral of the forward- and backward-propagating modes. A typical reflection spectrum of a uniform FBG is shown in Figure 1.4 [2].

Bragg grating reflectors are considered to be excellent temperature and strain sensing elements [26–28] because the measurements are an encoded wavelength. Due to this, the problems of intensity fluctuations that exist in many other types of fiber-based sensor systems are eliminated [2].

Because each Bragg grating reflector can be designated its own wavelength-encoded signature, a series of gratings can be written in the same fiber, each having a distinct Bragg resonance signal. Moreover, Bragg grating reflector has also proven to be a very useful element in tunable semiconductor laser [29–31], serving as one or both ends of the laser cavity, depending on the laser configuration. By changing the Bragg reflected feedback signal, the grating tunes the laser wavelength.

1.3.2   CHIRPED BRAGG GRATING

The refractive index profile of the grating can be modified to add other features, such as a linear, quadratic, or jump variation in the grating period. Therefore, the gratings have a non-uniform period along their length; this form is called a chirp, as shown in Figure 1.5. In a chirped FBG, the Bragg condition varies as a function of position along the grating. This is achieved by ensuring that the periodicity, Λ, varies as a function of position, or that the effective mode index, n_{eff}, varies as a function of position along the FBG, or through a combination of both. In this case, the reflected wavelength changes with the grating period, which results in a broadening in the reflected spectrum. A grating possessing a chirp has the property of adding dispersion, in which different wavelengths reflected from the grating will be subject to different delays [2, 3].

Chirped gratings have many important applications. Particularly, linearly chirped grating has found a special place in optics, such as a dispersion-correcting and compensating device [32, 33]. This application has also triggered the fabrication of

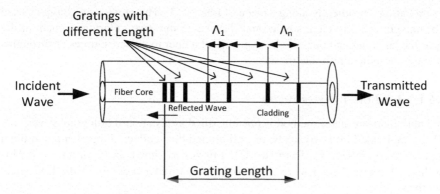

FIGURE 1.5 Chirped Bragg grating.

ultra-long, broad-bandwidth gratings of high quality, for high-bit-rate transmission [34] and in WDM transmission [35]. Some of the other applications include chirped pulse amplification [23], chirp compensation of gain-switched semiconductor lasers [27], sensing [26–29], higher-order fiber dispersion compensation [35], amplifier gain flattening [3], and band-pass filters [2].

1.3.3 TILTED BRAGG GRATING

In standard FBG, the gratings or the variations of the fiber core effective refractive index is along the length of the optical fiber axis and is typically uniform across the width of the fiber core. In contrast, in a tilted FBG, the variation of the fiber core effective refractive index is at an angle to the optical axis, as shown in Figure 1.6. The angle of tilt in tilted FBG has an effect on the reflected wavelength and bandwidth; the grating planes tilt and index modulation strength determine the coupling efficiency and the bandwidth of the light that is tapped out [2, 3].

The condition to satisfy the Bragg criteria of a tilted grating is similar to that of the uniform Bragg reflector [2]. Tilted Bragg grating has important applications in communication systems, such as flattening the gain spectrum of the erbium-doped fiber amplifier and mode conversion [3].

1.4 FABRICATION OF FIBER BRAGG GRATING

Fiber Bragg gratings are fabricated by techniques that are widely divided into two groups: holographic and non-interferometer, which are based on simple exposure to

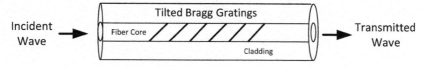

FIGURE 1.6 Tilted fiber Bragg grating structure.

UV radiation periodically along a piece of fiber [2, 3]. The previous techniques used a beam splitter to divide a single input UV beam into two, interfering them at the fiber; the latter depend on periodic exposure of a fiber to pulsed sources or through a spatially periodic amplitude mask.

1.4.1 THE BULK INTERFEROMETER

The bulk interferometer method is one classified as a standard holography, with the UV beam divided into two at a beam splitter and then collected together at a mutual angle of θ, by reflections from two UV mirrors as shown in Figure 1.7 [3]. This method allows the Bragg wavelength to be chosen independently of the UV wavelength as [3]

$$\lambda_B = \frac{n_{\text{eff}}\lambda_{\text{uv}}}{n_{\text{uv}}\sin(\theta/2)} \tag{1.1}$$

where n_{uv} is the refractive index of silica in the UV, λ_{uv} is the wavelength of the writing radiation, and θ is the mutual angle of the UV beams. This method was originally successfully used to write gratings at visible wavelengths [2, 3].

The interferometer is ideal for single-pulse writing of short gratings, and extreme care should be taken in the design of the optical mounts. Mechanical vibrations and the inherently long path lengths in air can cause the quality of the overlap to change over a period of time, limiting its application to short exposures. For low-coherence sources, the path difference between the two interfering beams must be equalized; a simple method is to introduce a mirror blank in one arm to compensate for the path imbalance imposed by the beam splitter, as shown in Figure 1.7 [3, 4].

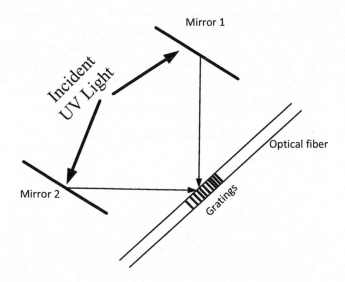

FIGURE 1.7 UV interferometer for writing Bragg gratings in optical fibers.

1.4.2 THE PHASE MASK

The phase mask method represents a major step toward easier fiber gratings fabrication, which is made possible by the application of it as a component of the interferometer. A phase mask is a relief grating etched in a silica plate [2, 3]. The significant characteristics of the phase mask are the grooves etched into a UV-transmitting silica mask plate, with a carefully controlled mark–space ratio as well as etch depth [3]. The principle of operation is based on the diffraction of an incident UV beam into several orders, m = 0, ±1, ±2. ... This is shown schematically in Figure 1.8. The incident and diffracted orders satisfy the general diffraction equation, with the period Λ_{pm} of the phase-mask given as [3, 4]

$$\Lambda_{pm} = \frac{m\lambda_{uv}}{\sin(\theta_m/2) - \sin\theta_i} \tag{1.2}$$

where $\theta_m/2$ is the angle of the diffracted order, l_{uv} the wavelength, and θ_i the angle of the incident UV beam. In the cases when the period of the grating lies between l_{uv} and $l_{uv}/2$, the incident wave is diffracted into only a single order (m = −1) with the rest of the power remaining in the transmitted wave (m = 0) [3].

When the UV beam radiation is incident at $\theta_i = 0$, the diffracted radiation is split into m = 0 and m = ±1 orders, as shown in Figure 1.9.

The interference pattern at the fiber of two such beams of orders ± 1 brought together by parallel mirrors has a period Λ_g related to the diffraction angle $\theta_m/2$ by [3, 4]

$$\Lambda_g = \frac{\lambda_{uv}}{2\sin(\theta_m/2)} = \frac{\Lambda_{pm}}{2} \tag{1.3}$$

The depth (d) of the etched fiber sections of the grating is a function of the UV beam wavelength, but the period is dependent only on the Bragg wavelength and the effective index of the mode. However, in the case of UV writing of gratings, it is necessary to ensure that the intensity of the transmitted zero-order beam is minimized and, ideally, blocked from arriving at the fiber [3, 4].

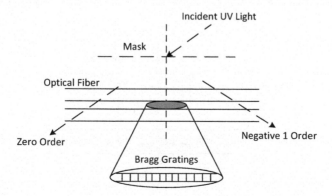

FIGURE 1.8   Schematic of the diffraction of an incident beam from a phase mask.

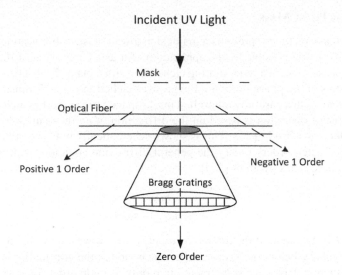

FIGURE 1.9 Normally incident UV beam diffracted into two ±1 orders.

1.4.3 Chirped Fiber Bragg Gratings

Chirped FBGs require a variation of the grating period Λ or a variation of the fiber effective refractive index n_{eff} along the length of the grating. Period chirped FBGs can be fabricated by bending the fiber with respect to the interferogram, as shown in Figure 1.10 [2], where the projection of the interference pattern onto the curved fiber creates a variation in the period. Bending the fiber creates a functional dependence of the grating period upon the radius of curvature, so that a linear or a quadratic chirp may be created.

Another technique for fabricating chirped FBGs is more flexible; it is capable of producing Bragg reflection with wide bandwidth, exploits the interference of beams with dissimilar wavefronts as shown in Figure 1.11 [4].

By introducing lenses of different focal length into the paths of the two beams in the holographic arrangement, the wavefront curvatures will differ at the fiber [2].

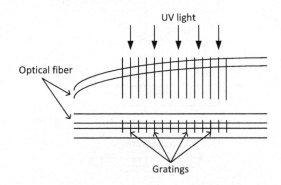

FIGURE 1.10 Writing a linearly chirped FBG by bending the optical fiber.

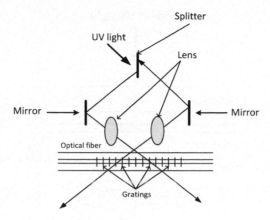

FIGURE 1.11   Using lens of different focus for writing chirped FBGs.

When the two beams are brought together to interfere, the resulting interferogram will no longer have constant period; the period varies as a function of distance along the axis of the fiber. Phase masks with constant period can also be used to fabricate chirped FBGs, as is shown in Figure 1.11 [2, 3]. By placing the fiber parallel to a constant period phase mask, a constant period is embossed into the core of the fiber. If the fiber is tilted, the period embossed is a function of the incident angle [2, 4]. The angle of incident of the collimated UV beam can be changed by the introduction of a lens, as shown in Figure 1.12 [4]. In this technique, a periodicity varying with grating length is produced and the inscribed chirp is determined based upon the mask's period, the inclined angle α, and lens characteristics.

The well-known phase mask technique is the simplest and easiest to use; however, it suffers from a lack of tunability of the Bragg wavelength when compared to the holographic method [2]. In addition to the common Bragg reflector, phase masks can be used to inscribe a continuously chirped period FBG [2, 3]. The chirp phase mask consists of a continuously varying mask period, as is shown in Figure 1.13 [2]. In this case, the writing process requires the fiber to be in close proximity to the phase mask but does not require that the fiber is tilted.

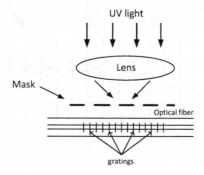

FIGURE 1.12   Using uniform phase mask to write a linear chirp FBG.

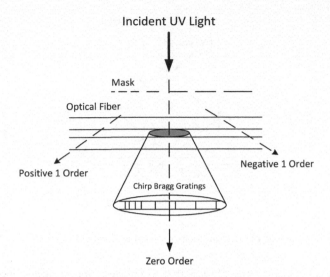

FIGURE 1.13   Using a chirped phase mask for writing a chirped FBG.

Previous sections have shown that a chirp period can be produced by making the effective refractive index vary periodically along the FBG. Furthermore, chirped FBGs can also be produced by making the effective refractive index of the propagation mode vary along the FBG [2, 3, 4, 27]. This can be realized by changing the guiding properties along the grating length, such as varying the diameter of the cladding of the fiber to a taper. This tapered fiber can be produced by differential etching by using a timed chemical etching method as shown in Figure 1.14. The taper can be produced by etching chemical technique [2, 3, 27] or by extending the fiber [36]. When a uniform periodic effective refractive mode index is writing in the fiber core of the tapered section, the chirped FBG can be created.

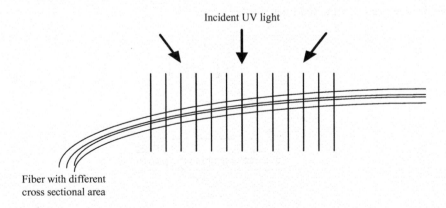

FIGURE 1.14   Using a tapered fiber to create a chirped FBG.

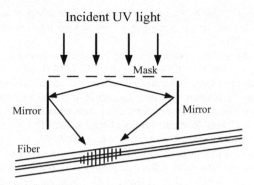

FIGURE 1.15   Method for writing slanted gratings.

1.4.4 SLANTED GRATING

If the fiber is tilted out of the plane of Figure 1.9, the grating inscribed in the fiber will be slanted in the direction of propagation of the mode [2, 3]. This, however, requires the interfering beams to have a large cross-sectional area so that the beams may overlap. However, this is inconvenient for most interferometers, since the cylindrical lens focuses the beams in the plane of the figure, unless the unfocused beam intensity is already high. An alternative and simple method for inscribing slanted gratings is to tilt the fiber in the plane of the figure, as shown in Figure 1.15 [3]. In this case, the coherence properties of the laser will determine the visibility of the fringes at the fiber. Since the fiber is at an angle to the incoming beams, the inscription of the grating depends on the overlap of the two beams [3].

1.5 SUMMARY

In this chapter, a brief review on the photosensitivity in optic fiber and an overview of the fundamental theory of fiber Bragg gratings and their development has been presented. The chapter also provided a review of the FBG types and the mechanisms for their fabrications.

REFERENCES

1. Hill, K.O., Fujii, Y., Johnson, D.C., and Kawasaki, B.S. Photosensitivity in optical fiber waveguides: Application to reflection filter fabrication. *Appl. Phys. Lett.* 1978, 32, 647–649.
2. Othonos, A., and Kalli, K. *Fiber Bragg Gratings-Fundamentals and Applications in Telecommunications and Sensing*, Artech House, Boston, MA, 1999.
3. Kashyap, R. *Fiber Bragg Gratings*, Academic Press, 2009.
4. Romero, R., Frazão, O., Marques, P.V.S., and Salgado, H.M. Ring chirped fibre Bragg grating for dynamic dispersion compensation. *Opt. Commun.* 2004, 242, 417–423.
5. Liu, H.Y., Liu, H.B., Peng, G.D., and Chu, P.L. Observation of type I and type II gratings behavior in polymer optical fiber. *Opt. Commun.* 2003, 220, 337–343.
6. Lin, D., Wang, L., and He, J.-J. Rate equation analysis of high speed Q-modulated semiconductor laser. *J. Lightwave Technol.* 2010, 28, 3128–3135.

7. Moon, D.S., Sun, G., Lin, A., Liu, X., and Chung, Y. Tunable dual-wavelength fiber laser based on a single fiber Bragg grating in a Sagnac loop interferometer. *Opt. Commun.* 2008, 281, 2513–2516.

8. Honzatko, P. All-optical wavelength converter based on fiber cross-phase modulation and fiber Bragg grating. *Opt. Commun.* 2010, 283, 1744–1749.

9. Biswas, P., Bandyopadhyay, S., Kesavan, K., Parivallal, S., Sundaram, B.A., Ravisankar, K., and Dasgupta, K. Investigation on packages of fiber Bragg grating for use as embeddable strain sensor in concrete structure. *Sens. Actuators A: Phys.* 2010, 157, 77–83.

10. Li, H., Ou, J., and Zhou, Z. Applications of optical fibre Bragg gratings sensing technology-based smart stay cables. *Opt. Lasers Eng.* 2009, 47, 1077–1084.

11. Payo, I., Feliu, V., and Cortázar, O.D. Fibre Bragg grating (FBG) sensor system for highly flexible single-link robots. *Sens. Actuators A: Phys.* 2009, 150, 24–39.

12. Nevers, D., Zhao, J., Sobolev, K., and Hanson, G. Investigation of strain-sensing materials based on EM surface wave propagation for steel bridge health monitoring. *Constr. Build. Mater.* 2011, 25, 3024–3029.

13. Cheng, W.H., Chiu, S.F., Hong, C.Y., and Chang, H.W. Spectral characteristics for a fiber grating external cavity laser. *Opt. Quant. Electron.* 2000, 32, 339–348.

14. Suhara, T. *Semiconductor Laser Fundamentals*, Marcel Dekker, New York, NY, 2004.

15. Timofeev, F.N., Simin, G.S., Shatalov, M., Gurevich, S., Bayvel, P., Wyatt, R., Lealman, I., and Kashyap, R. Experimental and theoretical study of high temperature-stability and low-chirp 1.55 μm semiconductor laser with an external fiber grating. *Fiber Integr. Opt.* 2000, 19, 327–353.

16. Wu, Z.M., Xia, G.Q., Deng, T., and He, Y.P. A theoretical model used to analyze the output characteristics of fiber grating external cavity semiconductor lasers. *Optik* 2009, 120, 136–140.

17. Zhou, H., Xia, G., Fan, Y., Deng, T., and Wu, Z. Output characteristics of weak-coupling fiber grating external cavity semiconductor laser. *Opto-Electron. Rev.* 2005, 13, 27–30.

18. Xia, G., wu, Z., and Zhou, H. Influence of external cavity length on lasing wavelength variation of fiber grating semiconductor laser with ambient temperature. *Optik* 2003, 114, 247–250.

19. Liu, Q., Ye, Q., Pan, Z., Luo, A., Cai, H., Qu, R., and Fang, Z. Synthesis of fiber Bragg grating for gain-narrowing compensation in high-power Nd: Glass chirped pulse amplification system. *Opt. Fiber Technol.* 2011, 17, 185–190.

20. Ugale, S.P., and Mishra, V. Modeling and characterization of fiber Bragg grating for maximum reflectivity. *Optik* 2011, 122, 1990–1993.

21. Ming, M., and Liu, K. *Principle and Applications of Optical Communication*, McGraw-Hill, New York, NY, 1996.

22. Park, S.-J., Lee, C.-J., Jeong, K.-T., Park, H.-J., Ahn, J.-G., and Song, K.-H. Fiber-to-the-home services based on wavelength-division-multiplexing passive optical network. *J. Lightwave Technol.* 2004, 22, 2581–2591.

23. Vázquez-Sánchez, R.A., Kuzin, E.A., García-Lara, C.M., May-Alarcón, M., Camas-Anzueto, J.L., Righini, G.C., and Miridonov, S.V. Radio-frequency interrogation of a fiber Bragg grating sensor in the configuration of a fiber laser with external cavities. *Optik* 2010, 121, 2040–2043.

24. Yeh, C.H., and Chi, S. Utilizations of fiber Bragg gratings and Fabry–Perot lasers for fast wavelength switching technique. *Opt. Commun.* 2005, 256, 73–77.

25. Lee, J.-R., Chong, S.Y., Yun, C.-Y., and Yoon, D.-J. A lasing wavelength stabilized simultaneous multipoint acoustic sensing system using pressure-coupled fiber Bragg gratings. *Opt. Lasers Eng.* 2011, 49, 110–120.

26. Yang, D.X., Yu, J.M., and Tao, X.M.Structural and mechanical properties of polymeric optical fiber. *Mater. Sci. Eng.* 2004, A364, 256–259.
27. Hill, K.O., and Meltz, G. Fiber Bragg grating technology fundamentals and overview. *J. Lightwave Technol.* 1997, 15, 1263–1276.
28. Han, J.-M., Baik, S.-J., Jeong, J.-Y., Im, K., Moon, H.-M., Noh, H.-R., and Choi, D.-S. Output characteristics of a simple FP-LD/FBG module. *Opt. Laser Technol.* 2007, 39, 313–316.
29. Liaw, S.-K., Hung, K.-L., Lin, Y.-T., Chiang, C.-C., and Shin, C.-S. C-band continuously tunable lasers using tunable fiber Bragg gratings. *Opt. Laser Technol.* 2007, 39, 1214–1217.
30. Yang, J., Tjin, S.C,, and Ngo, N.Q. Wideband tunable linear-cavity fiber laser source using strain-induced chirped fiber Bragg grating. *Opt. Laser Technol.* 2004, 36, 561–565.
31. Khijwania, S.K., Goh, C.S., Set, S.Y., and Kikuchi, K. A novel tunable dispersion slope compensator based on nonlinearly thermally chirped fiber Bragg grating. *Opt. Commun.* 2003, 227, 107–113.
32. Baskar, S., Suganthan, P.N., Ngo, N.Q., Alphones, A., and Zheng, R.T. Design of triangular FBG filter for sensor applications using covariance matrix adapted evolution algorithm. *Opt. Commun.* 2006, 260, 716–722.
33. Zhao, Y., and Zhou, C. Fast characterization of low-reflectance Bragg gratings in a polarization maintaining fiber using a reference grating. *Opt. Fiber Technol.* 2011, 17, 242–246.
34. Liang, S., Tjin, S.C., Ngo, N.Q., Zhang, C., and Li, L. Novel tunable fiber-optic edge filter based on modulating the chirp rate of a π-phase-shifted fiber Bragg grating in transmission. *Opt. Commun.* 2009, 282, 1363–1369.
35. Tan, Z., Yong, C., Liu, Y., Zhi, T., Jihong, C., Tigang, N., Kai, Z., Ting, C., and Jian, S. Cross-phase modulation in long-haul systems with chirped fiber Bragg gratings-based dispersion compensators. *Optik* 2007, 118, 216–220.
36. Mora, J., Villatoro, J., Díez, A., Cruz, J.L., and Andrés, M.V. Tuneable chirp in Bragg gratings written in tapered core fibers. *Opt. Commun.* 2002, 210, 51–55.
37. Deng, L., and Cai, C.S. Applications of fiber optic sensors in civil engineering. *Struct. Eng. Mech.* 2007, 25(5), 577–596.

2 Polymer Fiber Bragg Gratings

2.1 INTRODUCTION

With the development of the wavelength division multiplexed-passive optical network (WDM-PON) system for broadband, network security, and high-speed transmitted data capacity, fiber grating has become an indispensable device in high-performance optical communications systems due to its unique features such as wavelength selectivity, high tunabilty, and low-loss characteristics [1–4]. However, the application of silica optical fiber Bragg grating (SOF-BG) in flexible structures is limited due to some of the intrinsic physical properties of silica such as rigidness, brittleness, high Young's modulus, and low extensibility. Therefore, the change in the Bragg wavelength due to changes in temperature and strain is small [5, 6], which does not meet the requirements for WDM systems, since the expected bandwidth of these systems in the future will be more than 100 nm [7–11].

In the case of polymer optical fiber Bragg grating (POF-BG), the situation is totally different because the thermal effect and the strain sensitivity are much more than those of SOF-BG [2, 6]. For example, the Young's modulus for the polymer is $(0.1 \times 10^{10} \text{ N/m}^2)$ compared with $(7.13 \times 10^{10} \text{ N/m}^2)$ for silica; it is more than 70 times smaller [2, 12], which makes the mechanical tunable much better than that of SFO-BG. In addition, POF-BG has the merits of a negative and large thermo-optic effect; thereby, large refractive index tuning by heating can be obtained higher than for SOF-BG [2]. Consequently, high tuning range can be obtained easily by direct polymer gratings heating, with the consumption of very little power. Furthermore, the flexibility of the polymer gratings can make the tunability extend beyond the thermo-optic effect limitation [13, 14]. Based on the above consideration, POF-BGs have applications with which SOF-BGs cannot compete.

Table 2.1 gives a brief comparison between SOF-BG and POF-BG [2, 15].

It was demonstrated that the reflection wavelength of the POF-BG could be tuned over a wider range than those written in SOF-BG [2]. The large expansible and high thermo-optics coefficient of POF-BG results in an extensive sensitivity [7, 15], which makes it more suitable than SOF-BG in many applications, especially in sensor fields. In addition to the previous features, POF-BG is easy and flexible in photosensitive design, where many additives can be easily introduced into the fiber core during the fabrication process. Moreover, due to the bio-compatibility, POF-BG is an excellent candidate to be a sensor in medical applications [16, 17]. Currently, the poly(methylmethacrylate) (PMMA)-based POF is the most significant commercial POF product, due to its excellent optical properties and low cost [2, 15].

TABLE 2.1

Comparison between SOF and POF [2, 15]

Silica Optical Fiber (SOF)	Polymer Optical Fiber (POF)
Low attenuation	High attenuation
High operating temperature	Low operating temperature
High weight	Low weight
High production cost	Low production cost
Difficult to add dopants	Easy doping possibilities
Low NA and diameter	High NA and diameter

2.2 PROPERTIES OF POLYMER OPTICAL FIBER

As illustrated in the previous sections, a Bragg grating in a light-transmitting waveguide produces a very narrow band of reflected optical energy, with a maximum reflectivity at the characteristic wavelength of the grating as shown in Figure 1.11. Unlike the conventional glass fibers, the index of refraction in light-transmitting polymers typically varies inversely with the temperature [2, 15], leading to negative thermo-optic coefficients that are 10 to 30 times greater than the positive thermo-optic coefficient of conventional silica glass. This strong negative thermo-optical characteristic imbues POF-BGs, generally packaged on low thermal expansion substrates, with precise wavelength discrimination when used as tuning filters [2, 15].

The POF is characterized by high light absorption rates, as approximately 0.2 dB/cm compared with 0.2 dB/km for SOF at 1550 nm wavelength [2, 15]. However, a new polymer material has debuted in the laboratory: the perfluorinated polymer with light absorption as low as 0.3 dB/km at 1550 nm, close to that of SOF [18, 19]. Furthermore, considerable changes in temperature can be produced within POF and gratings due to the intrinsic self-heating. In addition to the temperature changes, this self-heating may cause undesirable shifts in the Bragg wavelength, which results in changes in the reflectivity/transmissivity of the grating response. Therefore, to facilitate the POF selection and the rational design of POF-BG, a complete understanding of the thermo-optic behavior of POF-BG should be obtained [2, 15, 21].

2.3 PHOTOSENSITIVITY IN POLYMER OPTICAL FIBER

The main technique to create Bragg gratings is through interaction with other light. By creating an interference pattern with prisms, mirrors, or a phase mask, this pattern can be written in the optical fiber [2]. But there is one condition: The optical fiber needs to be photosensitive. Photosensitivity means that the refractive index of a material changes when high-intensity light is applied. Therefore, the photosensitivity plays an important role in FBGs' performance [2, 15, 19]. Because POFs are only a recent development, almost all photosensitivity researches have been done on SOFs. The situation for POFs is, however, completely different. When UV light is applied, processes like polymerization of remaining monomers and cross-linking will change the refractive index. This makes their standard photosensitivity already larger, approximately 10^{-4} compared to 10^{-5} in un-doped SOF [2, 19]]. Furthermore, it is also considerably easier to increase the effect, and through the use of simple organic dopant bleaching techniques, refractive index changes around 10^{-2}, 10^{-1} can be obtained [2, 15, 19].

2.4 BRAGG GRATINGS IN POLYMER FIBERS

A Bragg grating acts as a filter that reflects light at the Bragg condition, which depends on the grating period. To make a tunable Bragg grating requires the ability to change the grating period. Since polymers have both a larger thermal expansion coefficient and are less stiff than glass fibers [8, 14, 15], it's possible to make temperature-or stress-tuned Bragg fibers with a large tuning range. The fabrication of gratings in POF relies on the photorefractive effect of the polymer, which differs from that of SOF. Practically, there are four ways to use light to change the refractive index of the polymer [2, 15]. However, the fabrication of POF-BG is similar to that for SOF-BG except for the difference in the irradiation wavelength. The phase mask method represents the way that is most used for Bragg grating fabrication in POF [2, 15, 19].

Bragg gratings in polymer fibers exhibit important advantages over those written in silica fibers, including pliability, large elongation, and high temperature sensitivity [2, 8, 10, 15]. These advantages are attributed to the intrinsic characteristics of polymer materials. Other advantages in the development of polymer FBGs originate from their feature of easy design in material system, and it is supported by two polymer optical fiber-related factors: low processing temperature and stronger material compatibility [2, 15].

A comparison of some important properties between PMMA and silica are listed in the Table 2.2 [2, 15]. The virtue of PMMA benefits its FBG performance either at its fabrication stage or in its application. When performing sensing functions in textile structural material or composites, POF-BGs have a unique advantage, because they are easily embedded or integrated in textile material [2, 15].

TABLE 2.2
Comparison of Physical Properties between Silica and PMMA [2, 15]

Properties	Silica	PMMA
Refractive index	1.45	1.48
Thermal-expansion coefficient (K^{-1})	4.1×10^{-7} at (25°C)	7×10^{-5} at (0–50°C)
Elongation break (%)	0.5~1.5	~6
Young's Modulus (N/m²)	7.13×10^{10}	0.1×10^{10}
Thermal-optic coefficient (K^{-1})	8.6×10^{-6}	-1×10^{-4}
Strain-optic coefficient (nm/milli-strain)	0.78	0.98

2.5 TUNING CHARACTERISTICS OF FIBER BRAGG GRATINGS

In this section, we intend to develop a mathematical model to describe the fiber Bragg grating that is used in this analysis. The model will represent the basis for the FBG design process and will allow the designer to estimate the expected characteristics of the design elements. For more clarity and continuity in the discussion on the FBG, it is better to begin with an overview of photosensitivity and mode propagation in optical fiber. This understanding of waveguide theory will lay the groundwork for the couple mode theory that is used in FBG analysis [15, 19].

2.5.1 PHOTOSENSITIVITY IN OPTICAL FIBER

As we mentioned, the photosensitivity in optical fiber usually refers to a permanent change in the effective index of refraction of the core fiber resulting from the exposure to UV light [2, 15, 19]. A good understanding of this phenomenon leads to absolute control on the characteristics of the refractive index mode fiber gratings (FGs) and consequently improves the Bragg grating (BG) performance. Because of the difference in the thermal expansion coefficients between the cladding and core regions, and also due to the non-uniform chemical reactions, the fiber fabrication process often ends up with non-ideal molecular structure. Due to these defects, multiple band absorptions within the UV light spectrum will appear [15, 19].

Essentially, the defects centers for the germanium (Ge(n)) are classified to be the largest contributor to the photo-induced refractive index changes in silica optical fiber, where these defects are the Ge sites that can trap electrons [19]. These electrons can be composed in the form of germanium oxygen deficient centers (GODCs). In this case, the Ge atom is directly bounded to another one due to the absence of an oxygen atom (wrong bond), that have been previously identified as Ge(0) and Ge(3). Other defects centers observed are Ge(1) and Ge(2) electron tarp centers, which take the form of four coordinated Ge atoms. These defects are the responsible on the absorption bands 242 nm and 330 nm, respectively [15, 19].

Moreover, drawing-induced defects are produced due to the large atomic displacement when the fiber glass cools on drawing; these defects are responsible on the 630 nm bands. In addition, there are other defects, including the non-bridging oxygen hole center (NBOHC) and the proxy radical (P-OHC), which are responsible on the absorption bands 160 nm and 260 nm, respectively [15, 19]. Figure 2.1 shows a germanium (Ge) or silica (Si) defects types of germanium-silicate optical fiber [15].

The rise to the change in the refractive index in SOF can be obtained by several mechanisms, such as photo-mechanical, photo-chemical, and thermo-chemical. The germanium-electron (GeE) hole trap centers are produced due to the bleaching of Ge-Ge/Ge-Si wrong bonds, which constitute the major absorption peak at 195 nm [15, 19].

This photo-chemical conversion of GODCs into GeE centers contributes significantly to the photo-induced refractive index changes in the germanium-silicate glass. Furthermore, the interaction of the absorbed photons with the glass will also give rise to alternation in the absorption spectrum. Consequently, this will change the refraction index at mode wavelength [15].

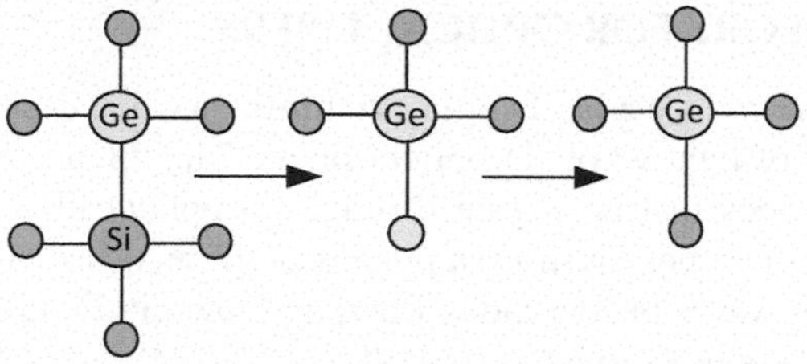

FIGURE 2.1 Types of Ge or Si in Germanium-Silica Optical Fiber [2].

For appreciable photosensitivity in optical fiber, certain core dopants are required. For example, the germanium-silicate demonstrates the most pronounced photosensitivity in a pure silica fiber. Photosensitivity in silica fibers increases with increasing germanium concentration [15, 19, 20].

In addition, the photosensitivity of the fiber core can be improved significantly by hydrogenation of the fiber before writing the gratings; the hydrogen molecules will interact with the host material, which leads to producing more GODC centers [21]. This technique has successfully caused phosphor-silicate fiber to demonstrate some photosensitivity. However, it produces excessive absorption loss in optical communication window at 1240 nm, 1390 nm, and 1410 nm due to molecular hydrogen, Si-OH, and Ge-OH, respectively [22].

On the other hand, polymer optical fibers are in most cases made of PMMA [15]; its chemical structure is shown in Figure 2.2. In order to change the refractive index, simpler techniques can be used compared to SOFs. Because the processing temperatures are much lower, organic dopants can be added during polymerization [5, 13, 22].

The refractive index of the POF core can be modulated by using the method of copolymerization of methylmethacrylate (MMA) monomer with other methacrylate monomers [7, 15, 23]. There are numerous methacrylate monomers which can contribute to the index modulation of core POF, either increasing or decreasing with different degree based on their chemical structure and the amount in the main chain of the copolymer [7]. The UV photosensitivity of core materials in PMMA-based POFs is changed by adding photoactive dopants or increasing the concentration of photosensitive co-monomer [19, 24]. For POFs, there are two types of dopants: benzophenone (BP) (Aldrich), a photo-initiator, and trans-4-tilbenemethanol (TS) (Aldrich), a photo-isomeric molecule [25].

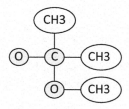

FIGURE 2.2 Chemical structure of PMMA [2].

2.6   MODE THEORY FOR OPTICAL FIBERS

In this section, we introduce a brief summary for theory that describes the propagation characteristics of light waves in optical fibers. This theory provides an understanding for the modes that propagate through an optical fiber when the light is
launched inside the core. For illustration purposes, the analysis will be limited to the
propagation of light waves in step-index fiber, as shown in Figure 2.3. The solution
for graded-index fibers is given by Jones [19].

For studying the light waves propagation in step-index fiber, we assume the optical
fiber with core radius a, and refraction index n_1 and n_2 for cladding and core, respectively. The indices of refraction are in the form $n_1 > n_2$. The transmitting light wave in
z-direction is described by Maxwell's equations in cylindrical coordinates [15, 19].
The wave propagation in z-direction is represented by the expression $e^{j(\omega t - \beta z)}$;
Maxwell's equations can be summed as [15, 19]

$$E_r = -\frac{j}{u^2}\left[\beta\frac{dE_z}{dr} + \omega\mu\frac{1}{r}\frac{dH}{d\theta}\right] \tag{2.1}$$

$$E_\theta = -\frac{j}{u^2}\left[\beta\frac{1}{r}\frac{dE_z}{d\theta} - \omega\mu\frac{dH_z}{dr}\right] \tag{2.2}$$

$$H_r = -\frac{j}{u^2}\left[\beta\frac{dH_z}{dr} - \omega\varepsilon\frac{1}{r}\frac{dH_z}{d\theta}\right] \tag{2.3}$$

$$H_\theta = -\frac{j}{u^2}\left[\beta\frac{1}{r}\frac{dH_z}{d\theta} + \omega\varepsilon\frac{dH_z}{dr}\right] \tag{2.4}$$

where

$$u^2 = \omega^2\mu\varepsilon - \beta^2 = k^2 - \beta^2 \tag{2.5}$$

E and H are the electric and magnetic field. μ and ε are the permeability and permittivity of the transmitted medium. The constant k is the free-space propagation and
the parameter β represented the z-component of the propagation vector determined
by the fields' interference at the core-cladding fiber shown in Figure 2.3, respectively
[19]. By using the above equations, the electric field wave equation in cylindrical
coordinates $E_z(r,\theta)$ can be written as [19]

$$\frac{d^2E_z}{dr^2} + \frac{1}{r}\frac{dE_z}{dr} + \frac{1}{r^2}\frac{d^2E_z}{d\theta^2} + u^2E_z = 0 \tag{2.6}$$

By assuming all terms with θ3B8 are periodically varying with a period $2\pi/v$, where
v is an integer value, Eq. (1.8) can be rewritten as [19]

$$\frac{d^2E_z}{dr^2} + \frac{1}{r}\frac{dE_z}{dr} + \left[u^2 - \frac{v^2}{r^2}\right]E_z = 0 \tag{2.7}$$

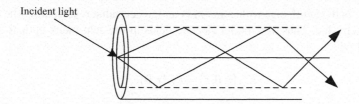

Incident light

FIGURE 2.3 Step-index fiber.

Equation (2.7) represents Bessel's differential equation, where its solutions are Bessel functions that vary with the parameter v. The solutions for the electric fields in the core and in the cladding as a result of the dependence on the radius r are given as [15, 19]

$$E_z(r,\theta) = AJ_v(ur)e^{jv\theta}r\langle a \qquad (2.8)$$

$$E_z(r,\theta) = BK_v(wr)e^{jv\theta}r\rangle a \qquad (2.9)$$

Similarly, the solutions in the core and in the cladding as a result of the dependence on the radius r for the magnetic fields are given as [20]

$$H_z(r,\theta) = CJ_v(ur)e^{jv\theta}r\langle a \qquad (2.10)$$

$$H_z(r,\theta) = DK_v(wr)e^{jv\theta}r\rangle a \qquad (2.11)$$

where $J_v(ur)$ and $K_v(wr)$ are the first and second kind of Bessel functions. The parameters u and w are given by [19]

$$u^2 = k_1^2 - \beta^2 \qquad (2.12)$$

and

$$w^2 = \beta^2 - k_2^2 \qquad (2.13)$$

where the free-space propagation constants k_1 and k_2 are defined by $2\pi n_1/\lambda$ and $2\pi n_2/\lambda$, respectively. If the boundary conditions are applied in Eqs. (2.10–2.13) for the E and H fields, the result is an eigenvalue equation that represents propagation parameters in the core and cladding [15]. The solutions to the eigenvalue equation will obtain the set of prorogation parameters that represent the possible modes of propagation. The only modes that can propagate inside the fiber are those that correspond to the eigenvalues. After successfully finding the eigenvalues, the propagation constant of the modes traveling inside the fiber can be calculated as [19]

$$\beta_{vm} = k_1^2 - u_{vm}^2 \qquad (2.14)$$

where u_{vm} is the vm-th eigenvalue and m is an integer value representing the number of the eigenvalue. Thus, the model electric fields in the step-index optical fiber are given as [19]

$$E_{vm}\left(r,\theta,z\right) = T_{vm}\left(r,\theta\right)e^{j\beta_{vm}z} \tag{2.15}$$

where $T_{vm}\left(r,\theta\right)$ is the amplitude of the transverse electric fields of the vm-th propagation constant [19].

2.7 FIBER MODES

Since the core of an optical fiber has a refraction index higher than the cladding [26, 27], all the light will be confined to the core when the total internal reflection condition is met [26, 27]. Essentially, the geometry design of the fiber determines the electromagnetic fields, or fiber modes, which can propagate in the fiber [11, 19, 26, 27].Generally, fiber modes can be classifications into two main categories, the radiation modes and the guided modes [19]. The radiation modes will carry the energy out of the core, where they are quickly dissipated [26, 27], while the guided modes are confined to the core, and propagate energy along the fiber, where the information and power are transporting [11, 19]. Based on the core size, the optical fiber can support many guided modes [26, 27], and each one has its own distinct velocity and can be further decomposed into orthogonal linearly polarized components and the modes transmitted represent the field distribution within the fiber [26, 27].

The type of the mode that propagates through the fiber depends on a few factors. These factors ultimately make up the fiber's V-number that determines which modes propagate in a fiber [11, 19, 26, 27]. This number gives indices of the core and the cladding refraction index, the core diameter, and the lasing wavelength. Thus, through fiber manufacturing, the type of the modes that are propagated are controlled.

2.7.1 OPTICAL FIBER PARAMETERS

2.7.1.1 Numerical Aperture (NA)

The Numerical Aperture (NA) of a fiber is defined as the sine of the largest angle an incident ray can have for total internal reflectance in the core (Figure 2.4). Rays launched outside the angle specified by a fiber's NA will excite radiation modes of the fiber [19, 26, 27].

$$NA = \sin\alpha = \sqrt{n_1^2 - n_2^2} = \sqrt{n_{core}^2 - n_{cladding}^2} \tag{2.16}$$

A higher core index, with respect to the cladding, means larger NA. A fiber's NA can be determined by measuring the divergence angle of the light cone it emits when all its modes are excited [11, 19, 26, 27].

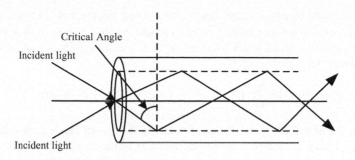

FIGURE 2.4 The physical meaning of NA.

2.7.1.2 V Number

The normalized frequency parameter of a fiber, also called the V number, is a useful specification given by [20, 31, 32]

$$V = \frac{2\pi NAa}{\lambda} = \frac{2\pi a}{\sqrt{n_{\text{core}}^2 - n_{\text{cladding}}^2}} \tag{2.17}$$

where a is the fiber core radius.

2.7.1.3 Cutoff Wavelength

The cutoff wavelength is the minimum wavelength in which a particular fiber still acts as a single-mode fiber given by [20, 31, 32]

$$\lambda_c = \frac{2\pi a}{2.405\sqrt{n_{\text{core}}^2 - n_{\text{cladding}}^2}} \tag{2.18}$$

When the propagate wavelength is above the λ_c, the fiber will be a single mode, while it becomes a multi-mode fiber if the wavelength is below the λ_c.

2.8 COUPLED-MODE THEORY FOR BRAGG GRATINGS

Coupled-mode theory is usually used to describe the relation between the spectral dependence of a fiber grating and the corresponding grating structure. While other techniques are available, we only consider coupled-mode theory due to unique properties; since it is straightforward, it is evident, and it accurately models the optical properties of fiber gratings of interest [19]. Coupled-mode theory represents the most appropriate tool to describe the light waves' propagation through a cylindrical waveguide with a slow change in the refractive index [15]. It provides a technique for obtaining quantitative information about the optical characteristics of FBG. Along this thesis, we assume that a single mode in the wavelength range and the fiber is lossless of interest. This means that we consider only one forward and one backward propagating mode. In addition, we assume that a weakly guiding fiber, i.e. the difference between the core and cladding refractive indices, is very small [19].

Coupled-mode theory assumes that the transverse electric field of perturbed fiber can be expressed as a superposition of ideal guided modes of unperturbed fiber [26]. Thus, the transverse electric fields can be expressed as a combination of forward and backward traveling waves as [26]

$$\vec{E}_T(r,\theta,z,t) = \sum_{vm}\left(A_{vm}(z)e^{j\beta_{vm}z} + B_{vm}(z)e^{-j\beta_{vm}z}\right).\vec{T}_{vmT}(r,\theta)e^{-j\omega t} \qquad (2.19)$$

where $A_{vm}(z)$ and $B_{vm}(z)$ are the amplitudes of the vm-th modes traveling in the forward and backward direction, respectively. ω is the angular frequency for the propagation modes and β is the propagation constant defined as $2\pi n_{eff}/\lambda$ where n_{eff} is the effective refractive index for a particular mode [19]. When a slowly varying perturbation appears in the fiber core, the traveling modes are forced to be coupled. Due to coupling with q-th mode, the amplitude of the m-th modes will vary along z-direction. As a result to coupling between the m-th mode and the q-th mode, the amplitudes A_m and B_m will vary according to [26]

$$\frac{dA_m}{dz} = j\sum_q A_q\left(K_{qm}^T + K_{qm}^z\right)e^{+j(\beta_q - \beta_m)z} + j\sum_q B_q\left(K_{qm}^T - K_{qm}^z\right)e^{-j(\beta_q + \beta_m)z} \quad (2.20)$$

$$\frac{dB_m}{dz} = -j\sum_q A_q\left(K_{qm}^T - K_{qm}^z\right)e^{+j(\beta_q + \beta_m)z} - j\sum_q B_q\left(K_{qm}^T + K_{qm}^z\right)e^{-j(\beta_q - \beta_m)z} \quad (2.21)$$

where K_{qm}^T and K_{qm}^z represent the transverse and longitudinal coupling coefficients between the modes m and q. In general, for optical fiber modes, the longitudinal coupling coefficient is usually neglected. The transverse coupling coefficient K_{qm}^T between the modes m and q is defined as [19, 26]

$$K_{qm}^T(z) = \frac{\omega}{4}\iint\limits_{2\pi,\infty} \Delta\varepsilon(r,\theta,z)\vec{T}_q^T(r,\theta).\vec{T}_m^{T*}(r,\theta)drd\theta \qquad (2.22)$$

where $\Delta\varepsilon(r,\theta,z)$ is the permittivity perturbation and * represents the complex conjugate. In common optical fiber, the changes in the refractive index due to the UV light are uniform inside the fiber core and neglected in the cladding. Under this assumption, the index change can be defined as [19, 26]

$$\delta n_{eff}(r,\theta,z) = \delta n_{eff}(z) = \overline{\delta n}_{eff}(z)\left\{1 + f\cos\left[\frac{2\pi}{\Lambda}z + \varphi(z)\right]\right\} \qquad (2.23)$$

where $\overline{\delta n}_{eff}(z)$ is the dc index change over a grating period, f is the fringe visibility of the index change, Λ is the grating period, and $\varphi(z)$ is the phase describing the grating chirp, respectively. Thus, the general equation for the transverses coupling coefficient K_{qm}^T is given by [19, 26]

$$K_{qm}^T(z) = \sigma_{qm}(z) + 2\kappa_{qm}(z)\cos\left[\frac{2\pi}{\Lambda}z + \varphi(z)\right] \qquad (2.24)$$

where $\sigma_{qm}(z)$ and $\kappa_{qm}(z)$ are the "dc" and "ac" coupling coefficients, respectively.

$$\sigma_{qm}(z) = \frac{\omega n_{core}}{2} \overline{\delta n}_{eff}(z) \iint_{core} \vec{T}_q^{T}(r,\theta).\vec{T}_m^{T*}(r,\theta)drd\theta \tag{2.25}$$

$$\kappa_{qm}(z) = \frac{f}{2}\sigma_{qm}(z) \tag{2.26}$$

2.9 MODELING OF FIBER BRAGG GRATINGS

In common FBGs, the coupling occurs between two identical modes that propagate in opposite directions on the z-axis. According to the fixed wavelength that is specified by the coupling coefficient, the interaction between the two modes will be dominated. As a result, for a mode of amplitude $A(z)$ and an identical counter-propagating mode of amplitude $B(z)$, the coupled-mode equations (2.20) and (2.21) are rewritten as [26]

$$\frac{dA(z)}{dz} = j\hat{\sigma}A(z)e^{j(\delta_d z - \varphi/2)} + j\kappa B(z)e^{(-j\delta_d z + \varphi/2)} \tag{2.27}$$

$$\frac{dB(z)}{dz} = -j\hat{\sigma}B(z)e^{(-j\delta_d z + \varphi/2)} - j\kappa^* A(z)e^{(j\delta_d z - \varphi/2)} \tag{2.28}$$

where $\hat{\sigma}$ is the general "dc" self-coupling coefficient defined as [15, 20, 31]

$$\hat{\sigma} = \delta_d + \sigma - \frac{1}{2}\frac{d\varphi}{dz} \tag{2.29}$$

The last term in Eq. (2.29) represents the possible chirp in the grating, δ_d is the detuning coefficient, which is always independed on z-variable, is defined as [15, 19, 26]

$$\delta_d = \beta - \frac{\pi}{\Lambda} = \beta - \beta_{Bragg} = 2\pi n_{eff}\left(\frac{1}{\lambda} - \frac{1}{\lambda_B}\right) \tag{2.30}$$

For single-mode fiber, the dc and ac coupling coefficients defined in Eqs. (2.28) and (2.29) can be simplified to [15, 19, 26]

$$\sigma = \frac{2\pi}{\lambda}\overline{\delta n}_{eff} \tag{2.31}$$

$$\kappa = \kappa^* = \frac{\pi}{\lambda}f\overline{\delta n}_{eff} \tag{2.32}$$

For a uniform grating along fiber axis, $\overline{\delta n}_{eff}$ is constant and $d\varphi/dz = 0$ [19]. These lead to a result that all the parameters κ, σ and $\hat{\sigma}$ are constants. Under these conditions the coupled-mode equations given in Eqs. (2.27) and (2.28) are simplified into first order differential equations with constant coefficients. The solution to these equations can be obtained by applying the suitable boundary conditions [19, 26].

2.10 SUMMARY

In this chapter, we have discussed the Bragg gratings in optical fiber-based polymer materials. A brief overview of the properties, the photosensitivity, and the fabrication of Bragg grating in polymer optical fiber has been presented. The tuning characteristics and the coupled-mode theory in polymer optical fibers were also discussed. The main parameters that determine the propagation modes through the fiber and the modeling of fiber Bragg grating were also reviewed.

REFERENCES

1. Park, S.-J., Lee, C.-J., Jeong, K.-T., Park, H.-J., Ahn, J.-G., and Song, K.-H. Fiber-to-the-home services based on wavelength-division-multiplexing passive optical network. *J. Lightwave Technol.* 2004, 22, 2581–2591.
2. Hisham, H.K., *Polymer Optical Fiber Bragg Gratings Technology: Spectral Response and Tuning Characteristics*, Lambert Academic Publishing, 2016.
3. Han, J.-M., Baik, S.-J., Jeong, J.-Y., Im, K., Moon, H.-M., Noh, H.-R., and Choi, D.-S. Output characteristics of a simple FP-LD/FBG module. *Opt. Laser Technol.* 2007, 39, 313–316.
4. Mora, J., Villatoro, J., Díez, A., Cruz, J.L., and Andrés, M.V. Tuneable chirp in Bragg gratings written in tapered core fibers. *Opt. Commun.* 2002, 210, 51–55.
5. Lin, D., Wang, L., and He, J.-J. Rate equation analysis of high speed Q-modulated semiconductor laser. *J. Lightwave Technol.* 2010, 28, 3128–3135.
6. Liu, Q., Ye, Q., Pan, Z., Luo, A., Cai, H., Qu, R., and Fang, Z. Synthesis of fiber Bragg grating for gain-narrowing compensation in high-power Nd: Glass chirped pulse amplification system. *Opt. Fiber Technol.* 2011, 17, 185–190.
7. Li, H., Ou, J., and Zhou, Z. Applications of optical fibre Bragg gratings sensing technology-based smart stay cables. *Opt. Lasers Eng.* 2009, 47, 1077–1084.
8. Timofeev, F.N., Simin, G.S., Shatalov, M., Gurevich, S., Bayvel, P., Wyatt, R., Lealman, I., and Kashyap, R. Experimental and theoretical study of high temperature-stability and low-chirp 1.55 μm semiconductor laser with an external fiber grating. *Fiber Integr. Opt.* 2000, 19, 327–353.
9. Yeh, C.H., and Chi, S. Utilizations of fiber Bragg gratings and Fabry–Perot lasers for fast wavelength switching technique. *Opt. Commun.* 2005, 256, 73–77.
10. Yang, D.X., Yu, J.M., and Tao, X.M. Structural and mechanical properties of polymeric optical fiber. *Mater. Sci. Eng.* 2004, A364, 256–259.
11. Hill, K.O., and Meltz, G. Fiber Bragg grating technology fundamentals and overview. *J. Lightwave Technol.* 1997, 15, 1263–1276.
12. Hisham, H.K. Bandwidth characteristics of FBG sensors for oil and gas applications. *Am. J. Sensor Technol.* 2017, 4, 30–34.
13. Moon, D.S., Sun, G., Lin, A., Liu, X., and Chung, Y. Tunable dual-wavelength fiber laser based on a single fiber Bragg grating in a Sagnac loop interferometer. *Opt. Commun.* 2008, 281, 2513–2516.
14. Suhara, T. *Semiconductor Laser Fundamentals*, Marcel Dekker, New York, 2004.
15. Kuzyk, M.G. *Polymer Optical Fiber: Materials, Physics, and Applications*, Taylor & Francis Group, 2007.

16. Zhao, Y., and Zhou, C. Fast characterization of low-reflectance Bragg gratings in a polarization maintaining fiber using a reference grating. *Opt. Fiber Technol.* 2011, 17, 242–246.
17. Tan, Z., Yong, C., Liu, Y., Zhi, T., Jihong, C., Tigang, N., Kai, Z., Ting, C., and Jian, S. Cross-phase modulation in long-haul systems with chirped fiber Bragg gratings-based dispersion compensators. *Optik* 2007, 118, 216–220.
18. Ugale, S.P., and Mishra, V. Modeling and characterization of fiber Bragg grating for maximum reflectivity. *Optik* 2011, 122, 1990–1993.
19. Othonos, A., and Kalli, K. *Fiber Bragg Gratings-Fundamentals and Applications in Telecommunications and Sensing*, Artech House, Boston, 1999.
20. Hisham, H.K. Numerical analysis of thermal dependence of the spectral response of polymer optical fiber Bragg grating. *Iraq J. Electr. Electron. Eng. (IJEEE)* 2016, 12.
21. Liu, H.Y., Liu, H.D., Peng, G.D., and Chu, P.L. Observation of type I and type II gratings behavior in polymer optical fiber. *Opt. Commun.* 2003, 220, 337–343.
22. Honzatko, P. All-optical wavelength converter based on fiber cross-phase modulation and fiber Bragg grating. *Opt. Commun.* 2010, 283, 1744–1749.
23. Biswas, P., Bandyopadhyay, S., Kesavan, K., Parivallal, S., Sundaram, B.A., Ravisankar, K., and Dasgupta, K. Investigation on packages of fiber Bragg grating for use as embeddable strain sensor in concrete structure. *Sens. Actuators A: Phys.* 2010, 157, 77–83.
24. Payo, I., Feliu, V., and Cortázar, O.D. Fibre Bragg grating (FBG) sensor system for highly flexible single-link robots. *Sens. Actuators A: Phys.* 2009, 150, 24–39.
25. Nevers, D., Zhao, J., Sobolev, K., and Hanson, G. Investigation of strain-sensing materials based on EM surface wave propagation for steel bridge health monitoring. *Constr. Build. Mater.* 2011, 25, 3024–3029.
26. Liu, M.M.-K. *Principles and Applications of Optical Communications*, Richard D. Iriwn Inc, Homewood, IL, 1996.
27. Senior, J.M. *Optical Fiber Communications Principles and Practice*, 3rd British Library, 2009.

3 Properties of Fiber Bragg Gratings

3.1 REFLECTIVITY OF GRATING FIBER

Consider a uniform fiber Bragg grating (FBG) with grating length L_g and forward-propagating wave incident from $z = -\infty$, and no backward-propagating wave for $z \geq L_g/2$. According to the model shown in Figure 3.1, the initial conditions for the coupled-mode equations are $\left(A(-L_g/2) = 1\right)$ and $\left(B(L_g/2) = 0\right)$ [1–3].

After imposing the aforementioned boundary conditions, the complex amplitude reflection coefficient ρ for uniform fiber Bragg grating can be obtained as [1]

$$\rho = \frac{B\left(-L_g/2\right)}{A\left(-L_g/2\right)} = \frac{-\kappa \sinh\left(\gamma_{\text{Bragg}} L_g\right)}{\hat{\sigma} \sinh\left(\gamma_{\text{Bragg}} L_g\right) + j\gamma_{\text{Bragg}} \cosh\left(\gamma_{\text{Bragg}} L_g\right)} \tag{3.1}$$

where γ_{Bragg} is the parameter related to the coupling coefficients defined as [1]

$$\gamma_{\text{Bragg}} = \sqrt{\kappa^2 - \hat{\sigma}^2} \tag{3.2}$$

The power reflection coefficient $R(\lambda)$ of a uniform fiber Bragg gratings can be defined as [1, 2]

$$R = |\rho|^2 = \frac{\sinh^2\left(\gamma_{\text{Bragg}} L_g\right)}{\cosh^2\left(\gamma_{\text{Bragg}} L_g\right) - \dfrac{\hat{\sigma}^2}{\kappa^2}} \tag{3.3}$$

According to the results given in Eq. (3.3), a number of an interesting features for fiber Bragg gratings can be obtained. From Eq. (3.3), the maximum reflectivity R_{max} for a Bragg grating is given by [1]

$$R_{\text{max}} = \tanh^2\left(\kappa L_g\right) \tag{3.4}$$

and it occurs at the condition $\hat{\sigma} = 0$ and at maximum wavelength λ_{max}, which is defined as [1]

$$\lambda_{\text{max}} = \left(1 + \frac{\overline{\delta n_{\text{eff}}}}{n_{\text{eff}}}\right)\lambda_B \tag{3.5}$$

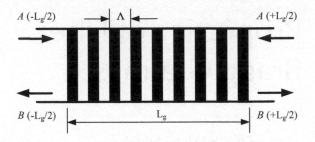

FIGURE 3.1   Uniform fiber Bragg gratings [3].

Inside the band gap (when $|\hat{\sigma}|\langle\kappa\rangle$), the amplitudes $A(z)$ and $B(z)$ are increased and decay exponentially along the fiber grating axis. In contrast, outside the band gap they develop sinusoidally. At the band edges, the fiber Bragg grating reflectivity is given by [1]

$$R_{\text{band-edge}} = \frac{\left(\kappa L_g\right)^2}{1+\left(\kappa L_g\right)^2} \tag{3.6}$$

The wavelengths at the band edges spectra are defined in terms of λ_{max} by the equation as [1]

$$\lambda_{\text{band-edge}} = \lambda_{\text{max}} \pm \frac{f\,\overline{\delta n}_{\text{eff}}}{2n_{\text{eff}}}.\lambda_B \tag{3.7}$$

Figure 3.2 shows the reflection and transmission spectral response for three different values of grating strength (κL_g) for silica optical fiber (SOF) and polymer optical fiber (POF) Bragg gratings (BGs), respectively [3]. The reflection and transmission peak values were obtained by adjusting the coupling coefficient in the coupled-mode equations. The gratings were designed to reflect light of wavelength $\lambda_B = 1550$ nm. As the result implies, by increasing κL_g, the peak reflectance value increases and the bandwidth of the Bragg reflector becomes narrower (i.e. longer gratings produce narrower spectral linewidths). This indicates that the bandwidth of the grating response can be tuned to a desired value by varying the grating strength. From the results in the figure, observe that the bandwidth for silica optical fiber for the same value of the grating strength is narrower than that for polymer optical fiber [3].

Figure 3.3 shows the time delay spectra response at room temperature for three different values of κL_g for SOF-BGs and POF-BGs, respectively [3]. Figure 3.3 also shows the reflectivity spectral response. Clearly, both reflectivity and time delay are symmetrical about the designed Bragg wavelength λ_B [3]. As the result implies, by increasing the κL_g value, the time delay decreases significantly and reaches its minimum value at the designed reflected wavelength for both silica and polymer optical fiber [3]. In addition, the time delay becomes appreciable near the band edges and side lobes of the reflection spectrum, where it tends to vary rapidly with wavelength.

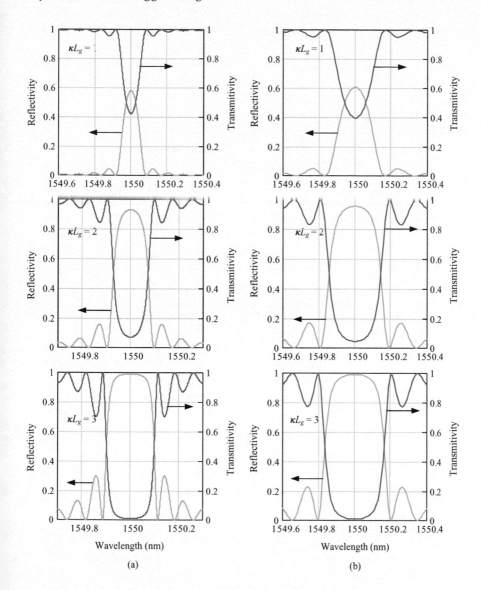

FIGURE 3.2 Reflection and transmission spectral response versus wavelength for a uniform Bragg grating with different values of grating strength for (a) silica optical fiber and (b) polymer optical fiber [3].

Moreover, results showed that the delay time for POF-BGs is less than that for SOF-BGs at the same value of κL_g. For example, when the grating strength is equal to $\kappa L_g = 1$, the delay time at the designed wavelength for POF-BGs is 18.7 ps compare with 73.6 ps for SOF-BGs. Also, when the grating strength increases to $\kappa L_g = 3$, the delay time for POF-BGs decreases to 10.1 ps compared with 32 ps for SOF-BGs [3].

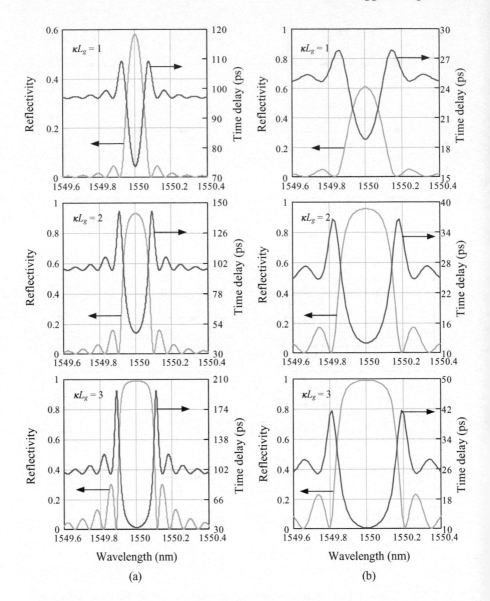

FIGURE 3.3 Reflection spectral response and delay time versus wavelength for uniform Bragg gratings with different values of grating strength for (a) silica optical fiber and (b) polymer optical fiber [3].

3.2 BANDWIDTH CHARACTERISTICS

The bandwidth for a uniform Bragg grating is the measures between the first zeros on either side of the maximum reflectivity. Based on Eq. (3.1), the normalized bandwidth of a Bragg grating measured at the band edges is defined as [1, 2]

$$\frac{\Delta\lambda_{\text{band-edge}}}{\lambda} = \frac{f\overline{\delta n}_{\text{eff}}}{n_{\text{eff}}} \tag{3.8}$$

Moreover, the normalized bandwidth $\left(\Delta\lambda_o/\lambda\right)$ for uniform Bragg grating based on the reflectivity definition given in Eq. (3.1) is defined by [1, 2]

$$\frac{\Delta\lambda_o}{\lambda} = \frac{f\overline{\delta n}_{\text{eff}}}{n_{\text{eff}}} \cdot \sqrt{1 + \left(\frac{\lambda_B}{f\overline{\delta n}_{\text{eff}}L_g}\right)^2} \tag{3.9}$$

For the case where the index of refraction change is weak (weak grating), $f\overline{\delta n}_{\text{eff}}$ is very small; thus, $f\overline{\delta n}_{\text{eff}} \langle\langle \dfrac{\lambda_B}{L_g}$ leads to [1]

$$\frac{\Delta\lambda_o}{\lambda} \rightarrow \frac{\lambda_B}{n_{\text{eff}}L_g} = \frac{2}{M} \tag{3.10}$$

where M is the total number of grating periods ($M = L_g/\Lambda$). In this case, the bandwidth is limited by the gratings length (length limited). However, in the case of strong gratings where $f\overline{\delta n}_{\text{eff}} \rangle\rangle \dfrac{\lambda_B}{L_g}$, the bandwidth becomes [1]

$$\frac{\Delta\lambda_o}{\lambda} \rightarrow \frac{f\overline{\delta n}_{\text{eff}}}{n_{\text{eff}}} \tag{3.11}$$

In the strong gratings, since the light does not penetrate the full length of the grating, the bandwidth is independent of grating length and directly proportional to the induced refractive index change. Therefore, for strong gratings, the bandwidth is similar, whether measured at the band edges, or as the full width at half-maximum (FWHM) [1, 2].

Figure 1 in an article by Hisham [2] shows the effect of refractive index change (δn) on the bandwidth characteristics of a uniform grating fiber. The obtained results in that article have observed that the 3-dB bandwidth increases with the increase in δn value, and this increment significantly increases by increasing the L_g value. The effect is approximately linear, especially with increasing the grating length (L_g), as shown in Figure 2 in that article [2]. These results show that the change in the fiber refractive index leads to a shift in the center Bragg wavelength, and this effect is used for calculating the external perturbations like temperature, strain, pressure, etc. [2].

Figure 3 in that article [2] has investigated the effect of grating length (L_g) on the reflectivity characteristics of a uniform obtained results show that the 3-dB bandwidth did not change significantly after L_g of 7 mm and is maintained at 1.0 nm for 10 mm [2]. Also, results obtained show that the 3-dB bandwidth has decreased exponentially with increasing L_g, and when the grating length was 7 mm, the 3-dB bandwidth was 1.0 nm and maintained subsequently for longer length [2].

These authors [2] have investigated the effect of refractive index change (δn) on the FWHM (i.e. the 3-dB bandwidth) characteristics of uniform FBG sensors for different L_g values. They found that the 3-dB bandwidth increases with the increase of δn value, and this increment significantly increases by increasing the L_g value. Also, they found that the change in the fiber refractive index leads to shift in the center Bragg wavelength, and this effect is used for calculating the external perturbations like temperature, strain, pressure, etc. [2].

3.3   DELAY AND DISPERSION CHARACTERISTICS

In addition to the properties of the reflectivity spectra, the group delay and the dispersion of the reflected light represent other spectral properties of interest in fiber Bragg gratings. These properties can be obtained from the phase of the complex amplitude reflection coefficient ρ defined in Eq. (3.1). The group delay τ_p of the reflected light defined as [1]

$$\tau_p = \frac{d\theta_p}{d\omega} = -\frac{\lambda^2}{2\pi c} \cdot \frac{d\theta_p}{d\lambda} \tag{3.12}$$

where θ_p is the phase of the complex amplitude reflection coefficient ρ. The dispersion of the reflected light D_p is defined as the rate of change of the group delay τ_p, with wavelength as follows [1]

$$D_p = \frac{d\tau_p}{d\lambda} = \frac{2\tau_p}{\lambda} - \frac{\lambda^2}{2\pi c} \cdot \frac{d^2\theta_p}{d\lambda^2} = -\frac{2\pi c}{\lambda^2} \cdot \frac{d^2\theta_p}{d\omega^2} \tag{3.13}$$

For making the grating appropriate for particular applications, the design parameters of the fiber Bragg grating can be adjusted accordingly to optimize the spectral properties [2]. Figure 3.4 shows the effect of coupling strength (ξ) on the dispersion properties of uniform POF Bragg grating [4]. The dotted line presents the dispersion response with the lowest value that can be calculated by the optimal coupling strength with genetic algorithms. The dashed line presents the dispersion response with the highest dispersion value that can be calculated by the coupling strength with the genetic algorithms [4]. The others curves shown in Figure 3.4 are the numerical simulation results with the coupling strength of 2.5/m, 10/m, 20/m, 30/m, 50/m, and 60/m, respectively [4].

From Figure 3.4, it is obvious that the POF Bragg grating with the lowest dispersion response obtained by the genetic algorithms has the highest ξ value of 60/m. Moreover, it is shown that the dispersion response fluctuates within a specified range with the change of the gratings length (L_g). In addition, Figure 3.4 shows high consensus between the results that obtained by numerical analysis and that obtained by the genetic algorithms [4].

Moreover, in Figure 3.5, the dispersion responses range from approximately zero ps/nm (i.e. at the minimum value of grating length [i.e. $L_g \leq 2$ mm]) to several

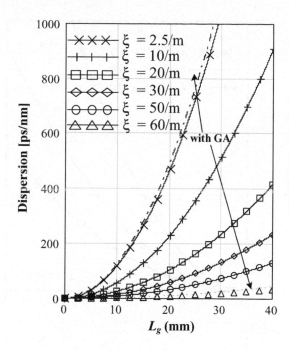

FIGURE 3.4 Dispersion properties of uniform POF function to gratings length (L_g) at different ξ value [4].

thousand ps/nm at high grating length (i.e. $L_g \geq 30$ mm) based on the value of ξ [5]. However, as is well known, the minimum value of grating length leads to reduced effective reflectivity for the gratings fiber and then the performance of the system [4]. Therefore, we are forced to work with high grating length to ensure high reflectivity operating. So, this leads to an increase in the dispersion based on the ξ value [5]. Also, the results obtained by the numerical analysis show that by increasing the ξ value from 2.5/m to 60/m, the dispersion responses are reduced significantly with L_g, and these results are totally consistent with those calculated by the genetic algorithms [4]. This is due to increased coupling strength between the forward and backward waves which leads to increased grating reflectivity and then reduces the dispersion value [4].

One of important parameters that controls the reflectivity and the spectral shape of the fiber grating is the grating's average refractive index change (δ_n) [1, 4] where it may cause shift in the center wavelength, assuming that the δ distributes uniformly across the fiber core and is nonexistent in the outside region, which is often true due to the lack of photosensitivity of the cladding [4].

Figures 3.6 and 3.7 show numerically the dispersion responses based on the optimal values for the ξ that are calculated by the genetic algorithms at different δn values [4]. From Figures 3.6 and 3.7, it is obvious that the dispersion responses are affected by the value of δ_n. This can be understood as: by reducing δ_n to a small value ($\delta_n \rightarrow 0$), the wavelength is shifted to a region near their Bragg condition, which results in increment in the grating reflectivity and a reduction in the dispersion value [1, 5].

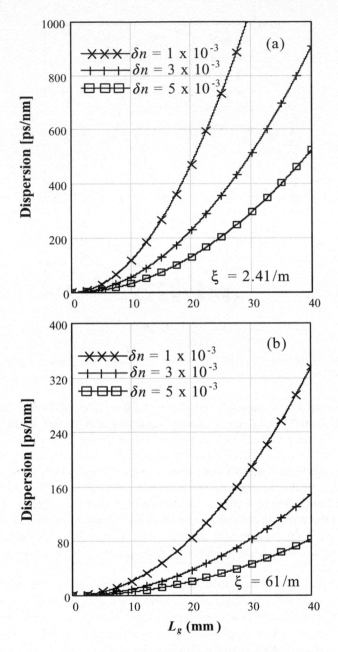

FIGURE 3.5 Effect of average refractive index (δn) on dispersion properties of uniform POF function to grating length (L_g) for (a) $\xi = 2.41/m$ and (b) $\xi = 61/m$ [4].

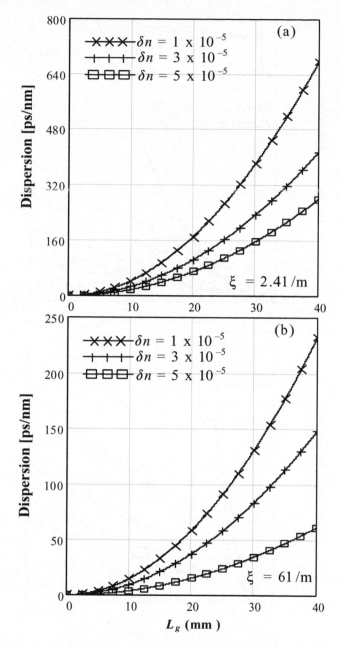

FIGURE 3.6 Effect of average refractive index (δn) on dispersion properties of uniform POF function to gratings length (L_g) based on the optimal values of (a) ξ =2.41/m and (b) ξ = 61/m.

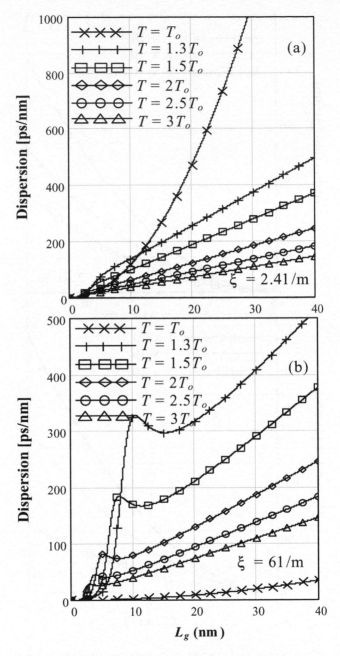

FIGURE 3.7   Effect of temperature (T) variation on dispersion properties of uniform POF function to gratings length (L_g) (a) $\xi = 2.41/m$ and (b) $\xi = 61/m$ [4].

It is very important that at the same ξ value, the dispersion amplitude in Figure 3.6, is less than that of Figure 3.7 [4].

Figure 3.8 shows the effect of temperature (T) variation on the dispersion properties [4]. Results in Figure 3.8 (a) and (b) are obtained numerically based on the optimal values for the ξ that are calculated by the genetic algorithms [4]. In this analysis, T has varied from T_o to 3 T_o. As shown, the change in dispersion response with

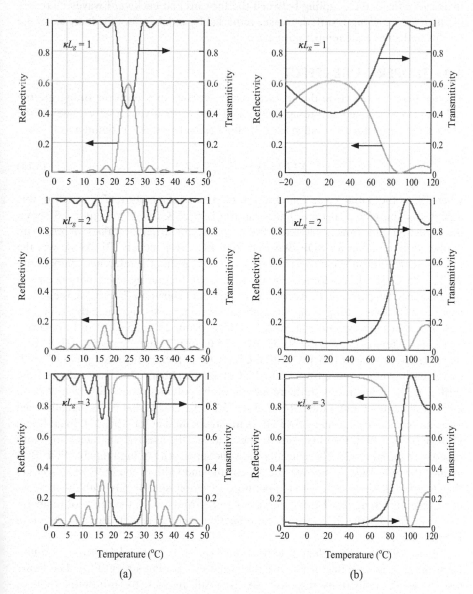

FIGURE 3.8 Reflection and transmission spectral versus temperature variation for a uniform Bragg grating with different values of grating strength for (a) silica optical fiber and (b) polymer optical fiber [3].

temperature in Figure 3.6a [4] is more linear than that in (b), while the value in (b) is less than that in (a). This change is basically due to the temperature dependence for the refractive index of the fiber grating [3–5] which, by shifting temperature from its reference value, the refractive index is changed and leads to a shift in the Bragg condition to another wavelength [3–6]. However, at the high ξ value [4], the dispersion properties with temperature variation are less than that at the low ξ, where, with the increase of the ξ, the coupling between the forward and backward waves increases [4], then the wavelength shift decreases remarkably, and therefore the increase of the ξ helps to decrease the temperature effect [4].

3.4   TEMPERATURE EFFECT

The temperature effect on the spectral characteristics of a uniform Bragg grating reflector is investigated according to its effect on the effective refractive index of the fiber. The temperature dependence of the fiber refractive index is defined as [6–9]

$$X(T) = X_o + \frac{\partial X}{\partial T}(T - T_o) \tag{3.14}$$

where X_o is the initial value found at the reference temperature (T_o), which in this analysis is considered at room temperature $(T_o = 25°C)$ [7–10]. Figure 3.8 shows the reflection and transmission spectral response for different values of κL_g with temperature variation for SOF-BGs and POF-BGs, respectively [3]. As shown, the reflection and transmission spectra are symmetric around the reference temperature T_o $(T_o = 25°C)$ [1, 3]. In addition, the peak value of the grating reflectivity occurs at the reference temperature T_o. According to the result shown in Figure 3.8, the reflectivity of SOF-BGs with grating strength $\kappa L_g = 1$ decreases significantly from 58% to 0.05% by changing temperature $\Delta T = 10°$ C (from 25 to 35°C) [3]. In contrast, by changing temperature $\Delta T = 50°$ C (from 25 to 75°C), the reflectivity of POF-BGs decreases from 60% to 15% [3]. With the increase of the grating strength to $\kappa L_g = 3$, by increasing temperature from 25 to 35°C, the silica fiber reflectivity reduces from 99% to 6.5%, compared with the reduction in the polymer reflectivity from 99% to 89% by changing temperature from 25 to 75°C [3]. This is due to the negative and large thermo-optic coefficient for POF compared with that for silica [5]. In addition, POF-BGs show high stability with temperature for the spectral characteristic compared with the silica [3]. These results show the superiority of high tunable POF-BGs against the SOF-BGs. Moreover, the spectrum bandwidth of the POF-BGs is larger than that for the silica fiber with temperature variation, specifically with the increase of the grating strength, where by increasing κL_g from 1 to 3, the range of temperature operation for the first zero of the reflection spectral is increased, as shown in Figure 3.8 [3].

Figure 3.9 shows the effect of temperature variation on the delay time for a uniform Bragg grating reflector for different values of grating strength [3]. The figure also shows the reflectivity response. As the result implies, the minimum value of the delay time for the light reflected back from the grating occurs at the reference temperature T_o. In addition, the temperature operation range for low time delay in POF-BGs is greater than that for SOF-BGs [3].

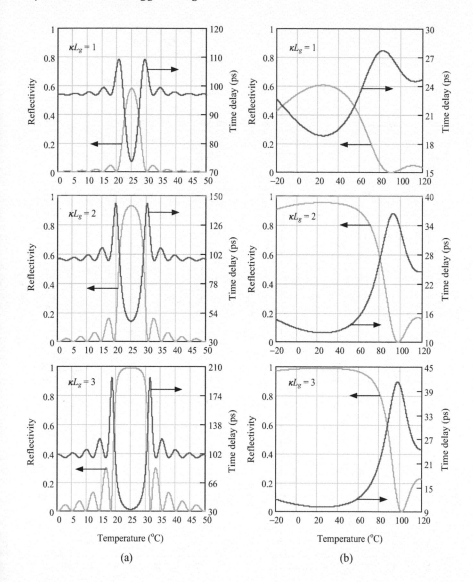

FIGURE 3.9 Reflection spectral and delay time versus temperature variation for a uniform Bragg grating with different values of grating strength for (a) silica optical fiber and (b) polymer optical fiber [3].

Figure 3.10 and 3.11 show the effect of temperature variation on the spectral response for SOF-BGs and POF-BGs for two different values of the fiber index modulation, respectively [3]. Although the range of temperature operation is large, the grating length is equal to 1 mm, as shown in Figure 3.10; however, the peak value of the grating reflectivity is very low, around 4% [3]. In contrast, the peak reflectivity value increases to around 60% when the index modulation value increases to 5 × 10^{-4}, as in Figure 3.6. In addition, the grating bandwidth is length limited (i.e. weak

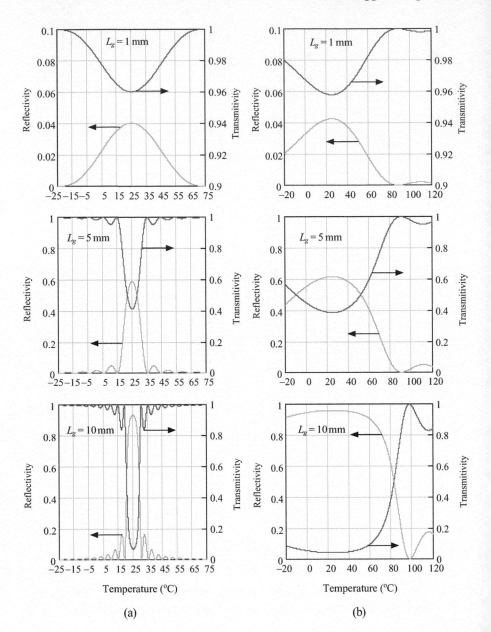

FIGURE 3.10 Reflection and transmission spectral versus temperature variation for a uniform Bragg grating with index modulation $\Delta n = 1 \times 10^{-4}$ for different values of grating length (a) silica optical fiber and (b) polymer optical fiber [3].

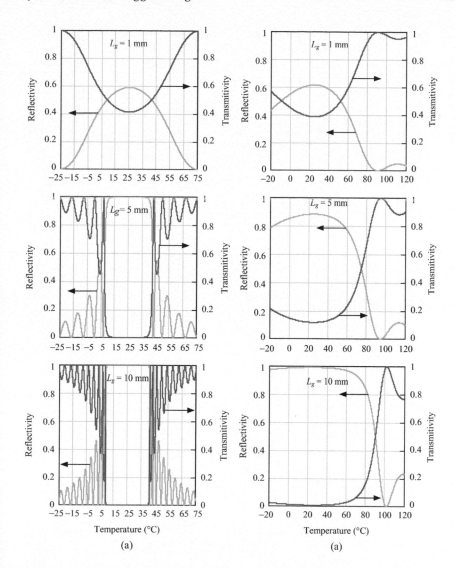

FIGURE 3.11 Reflection and transmission spectral versus temperature variation for a uniform Bragg grating with index modulation $\Delta n = 5\times10^{-4}$ for different values of grating length (a) silica optical fiber and (b) polymer optical fiber [3].

grating [1, 5]), specifically more for the silica fiber, where with the increase of the grating length, the grating bandwidth is changed [3]. In contrast, when the grating length increases to 5 mm and 10 mm with index modulation value equal to 5×10^{-4}, the silica fiber Bragg grating bandwidth becomes length independent (i.e. strong grating) [3].

Figures 3.12 and 3.13 show the effect of temperature variation on the time delay for silica and polymer fiber grating with index modulation value equal to 1×10^{-4}

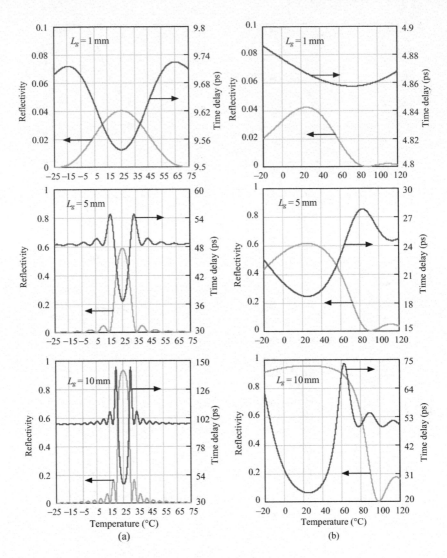

FIGURE 3.12 Reflection spectral and time delay versus temperature variation for a uniform Bragg grating with index modulation $\Delta n = 1 \times 10^{-4}$ for different values of grating length (a) silica optical fiber and (b) polymer optical fiber [3].

and 5×10^{-4}, respectively [3]. It is observed that for 1×10^{-4} index modulation, by increasing the grating length from 1 mm to 10 mm, the peak delay time increases. In contrast, for 5×10^{-4} index modulation, the increase in the peak value of the time delay with grating length is reduced [3]. It is clear from the results given in Figures 3.4 through 3.9, that the temperature operation range for high reflectivity and lower time delay for a polymer optical fiber with a uniform Bragg grating is greater than that for silica optical fiber [3].

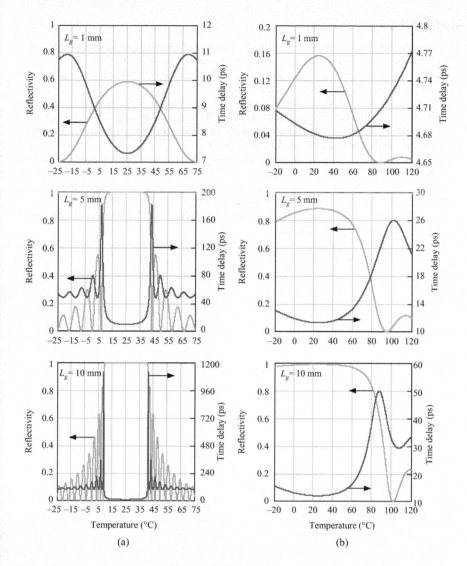

FIGURE 3.13 Reflection spectra and time delay versus temperature variation for a uniform Bragg grating with index modulation $\Delta n = 5 \times 10^{-4}$ for different values of grating length (a) silica optical fiber and (b) polymer optical fiber [3].

3.5 SUMMARY

In the chapter, comprehensive reviews of the properties of silica and polymer fiber Bragg gratings were discussed intensively and in ways that are contrary to those in previous studies. The effects of the grating length, fiber refractive index, coupling strength, and temperature variation on reflective, transmission, bandwidth, turn on time delay, and dispersion responses have been investigated.

REFERENCES

1. Othonos, A., and Kalli, K. *Fiber Bragg Gratings: Fundamentals and Applications in Telecommunications and Sensing*, Artech House, Norwood, MA, 1999.
2. Hisham, H.K. Bandwidth characteristics of FBG sensors for oil and gas applications. *Am. J. Sensor Technol.* 2017, 4 (1), 30–34.
3. Hisham, H.K. Numerical analysis of thermal dependence of the spectral response of polymer optical fiber Bragg gratings. *Iraq J. Electr. Electron. Eng.* 2016, 12 (1), 85–95.
4. Hisham, H.K. Low dispersion performance of plastic fiber grating using genetic algorithms. *Al-Nahrain J. Eng. Sci.* In Press, 21 (1).
5. Kuzyk, M.G. *Polymer Optical Fiber: Materials, Physics, and Applications*, CRC Press, Boca Raton, FL, 2007.
6. Hisham, H.K. Effect of temperature variations on strain response of polymer Bragg grating optical fibers. *Iraq J. Electr. Electron. Eng.* 2017, 13, 53–58.
7. Hisham, H.K. Design methodology for reducing RIN level in DFB lasers. *Iraq J. Electr. Electron. Eng.* 2016, 12, 207–213.
8. Hisham, H.K. Turn_on time reduction in VCSELs by optimizing laser parameters. *Iraq J. Electr. Electron. Eng.* 2016, 12.

4 Sensor Mechanisms in Fiber Bragg Gratings

4.1 SIMULATION METHODS OF FIBER BRAGG GRATINGS

In the previous chapters, we have provided the coupled-mode theory that determines the significant parameters for fiber Bragg gratings (FBG) and related them to the spectral properties. To calculate reflection, transmission, delay, and dispersion characteristics of the FBG, the coupled-mode equations must be solved [1–3]. There are several techniques can be used for this purpose. These techniques differ in the complexity, accuracy, and speed of computation [1, 2]. Therefore, the requirements from the designer make a compromise between them for optimum choice.

4.1.1 DIRECT NUMERICAL INTEGRATION

The first and most evident technique is the direct numerical integration method of the coupled-mode equations [1, 2]. In this approach, the coupled equations are directly integrated by using one of the numerical methods after applying the boundary conditions; we can start at the end of the grating $z = L_g$ and integrate backward to $z = 0$ [1] (Figure 3.1). Although the direct numerical integration has the inherent advantage of accuracy, it lacks in computation speed [1].

4.1.2 TRANSFER MATRIX METHOD

The principle behind the transfer matrix method is to divide the grating structure into a multiple number M of uniform sections [1, 2] so that each section can be approximately treated as uniform grating, and then identify each one by using a (2×2) matrix [1]. The information contained in each matrix is related to a specific section. Then, the individual matrices are successively gathered by multiplying them along the grating length to find the total behavior for the entire grating [1, 2].

The first step in this technique is to divide the grating structure into M uniform matrix components with A_k and B_k being the field amplitudes after transverse section k. From the derivation of Eq. (4.1), the boundary conditions give the starting point $A_o = A(+L_g/2) = 1$ and $B_o = (-L_g/2) = 0$. The propagation of each section k is described by a transfer matrix T_k expressed as [1, 2]

$$\begin{pmatrix} A_k \\ B_k \end{pmatrix} = T_k \begin{pmatrix} A_{k-1} \\ B_{k-1} \end{pmatrix} \tag{4.1}$$

where the transfer matrix T_k is given as [1, 2]

$$T_k = \left(\begin{array}{c} \cosh\left(\gamma_{\text{Bragg}}\Delta z\right) - j\dfrac{\hat{\sigma}}{\gamma_{\text{Bragg}}}\sinh\left(\gamma_{\text{Bragg}}\Delta z\right) \\[4mm] + j\dfrac{\kappa}{\gamma_{\text{Bragg}}}\sinh\left(\gamma_{\text{Bragg}}\Delta z\right) \\[6mm] -j\dfrac{\kappa}{\gamma_{\text{Bragg}}}\sinh\left(\gamma_{\text{Bragg}}\Delta z\right) \\[4mm] \cosh\left(\gamma_{\text{Bragg}}\Delta z\right) + j\dfrac{\hat{\sigma}}{\gamma_{\text{Bragg}}}\sinh\left(\gamma_{\text{Bragg}}\Delta z\right) \end{array} \right) \tag{4.2}$$

where Δz is the length of the k-th uniform section. When the matrices of all the individual sections are known, the output amplitudes can be calculated as [1, 2]

$$\begin{bmatrix} A_M \\ B_M \end{bmatrix} = T_M \cdot T_{M-1} \ldots\ldots T_k \ldots\ldots T_1 \cdot \begin{bmatrix} A_o \\ B_o \end{bmatrix} \tag{4.3}$$

The accuracy of this method depends on the number of uniform section M that are used in the analysis. Larger numbers of M are used; a higher accuracy is obtained. However, M cannot be increased arbitrarily. The choice of M should be satisfy the condition that the length of each uniform section dz must be larger than the grating period $(\Delta z \rangle \Lambda)$. In other words, M must satisfy [1, 2]

$$M \langle \frac{2n_{\text{eff}}L_g}{\lambda_B} \tag{4.4}$$

The biggest advantage of this method is the possibility to apply it for both uniform and non-uniform grating such as chirped and apodized gratings [3]. Figure 4.1 shows how the transfer matrix method is applied to the uniform and non-uniform gratings. In addition, the accuracy and computation speed of this method can be adjusted or controlled in accordance with the design requirements [3].

4.1.3 DISCRETIZED GRATING METHOD

Instead of dividing the grating structure into multiple numbers of uniform sections as above, we can discretize such that the whole grating becomes a pile of complex, discrete reflectors. Then, each of the transfer matrices above is replaced by $T^{\Delta z} \times T_k^{\zeta}$, where $T^{\Delta z}$ and T_k^{ζ} are the pure propagation and discrete reflector matrices, respectively, are defined as [1, 2]

$$T^{\Delta z} = \begin{bmatrix} \exp\left(j\hat{\sigma}\Delta z\right) & 0 \\ 0 & \exp\left(-j\hat{\sigma}\Delta z\right) \end{bmatrix} \tag{4.5}$$

$$T_k^{\zeta} = \left(1 - |\zeta_k|^2\right)^{-1/2} \cdot \begin{bmatrix} 1 & -\zeta_k^* \\ -\zeta_k & 1 \end{bmatrix} \tag{4.6}$$

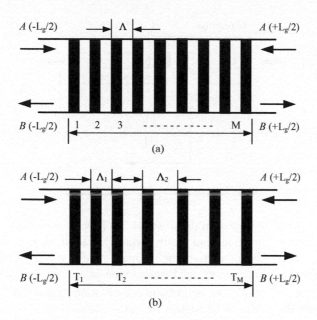

FIGURE 4.1   Using Transfer Matrix Method (a) uniform Bragg grating and (b) non-uniform Bragg grating.

where ζ_k is the discrete reflection coefficient given by [1]

$$\zeta_k = -\tanh\left(\frac{|\kappa_k|.\Delta z.\kappa_k^*}{|\kappa_k|}\right) \tag{4.7}$$

It is evident to show that the task of transferring the fields using $T^{\Delta z} \times T_k^{\zeta}$, similar to Eq. (3.1), ρ is given by [1, 2]

$$\rho(z;\hat{\sigma}) = \frac{\zeta_k + \rho(z + \Delta z;\hat{\sigma})\exp(2j\hat{\sigma}\Delta z)}{1 + \zeta_k^*\rho(z + \Delta z;\hat{\sigma})\exp(2j\hat{\sigma}\Delta z)} \tag{4.8}$$

According to the discretized method [1], the reflection coefficient ρ of the grating is obtained by setting $\rho(L_g;\hat{\sigma}) = 0$ and then working Eq. (4.1) backward to $z = 0$, yielding the spectrum $\rho(\hat{\sigma}) = \rho(0;\hat{\sigma})$ [1, 2]. In contrast to the direct numerical integration technique, the propagation of $\rho(z;\hat{\sigma})$ is exact for the discrete structure model, and the numerical stability is therefore as good as for the transfer matrix method [1, 2].

4.2   FIBER BRAGG GRATING SENSOR

As we explained in the previous sections, the main work of the FBGs is a wave-length-selective filter/reflector, which is obtained by introducing a periodic refractive index structure within the core of an optical fiber, as shown in Figure 4.2 [1–3].

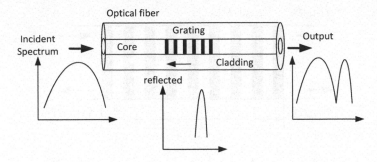

FIGURE 4.2   A uniform fiber Bragg grating.

In appropriate circumstances, if the wavelets that have been created in each plane are in phase, certain directions can be discovered. In addition, if these directions are consistent with the fibers' pattern, the resonant condition is satisfied and strong scattering will occur. In this case, the forward momentum is reflected back by the grating momentum $\vec{K}$ [1, 2]. The momentum conservation requires that the incident wave-vector plus the grating wave-vectors equal the scattered wave-vector. This simply stated as [1, 2]

$$\vec{k}_2 = \vec{k}_1 + \vec{K} \tag{4.9}$$

where $\vec{k}_1$, $\vec{k}_2$ are the forward and backward wave-vectors' propagation modes, and $\vec{K} = (2\pi/\Lambda)$ is the grating momentum, respectively [1]. Since the forward and backward propagating waves are identical in terms of photon frequencies, this leads that $\lambda_1 = \lambda_2 = \lambda_B$, where λ_B is the light wavelength that satisfies Bragg condition [2]. From Eq. (4.9) we obtain [1]

$$2\left(\frac{2\pi n_{\text{eff}}}{\lambda_B}\right) = \left(\frac{2\pi}{\Lambda}\right) \tag{4.10}$$

The equation above leads to the first order Bragg condition [1, 2]

$$\lambda_B = 2n_{\text{eff}}\Lambda \tag{4.11}$$

Thus, the incident light on a Bragg grating with a grating period Λ and effective refractive index mode n_{eff} will be reflected back at the wavelength λ_B typically named Bragg wavelength [1, 2]. The Bragg wavelength is affected significantly by the change of the ambient temperature, applied strain, and incident wavelength [1, 2]. Based on Eq. (4.11), the difference shift in Bragg wavelength which results from a change in these parameters is obtained as [1]

$$\Delta\lambda = 2\left(\Lambda\frac{\partial n_{\text{eff}}}{\partial L_g} + n_{\text{eff}}\frac{\partial\Lambda}{\partial L_g}\right)\Delta L_g + 2\left(\Lambda\frac{\partial n_{\text{eff}}}{\partial T} + n_{\text{eff}}\frac{\partial\Lambda}{\partial T}\right)\Delta T$$

$$+2\left(\Lambda\frac{\partial n_{\text{eff}}}{\partial\lambda}+n_{\text{eff}}\frac{\partial\Lambda}{\partial\lambda}\right)\Delta\lambda \qquad (4.12)$$

The variation in refractive index due to a change in incident wavelength can be neglected [1, 2, 4, 5, 6], which leads to reduce Eq. (4.12) to

$$\Delta\lambda = 2\left(\Lambda\frac{\partial n_{\text{eff}}}{\partial L_g}+n_{\text{eff}}\frac{\partial\Lambda}{\partial L_g}\right)\Delta L_g + 2\left(\Lambda\frac{\partial n_{\text{eff}}}{\partial T}+n_{\text{eff}}\frac{\partial\Lambda}{\partial T}\right)\Delta T \qquad (4.13)$$

where ΔL_g and ΔT are the change in grating length due to applied strain and the change in ambient temperature, respectively. We can separate Eq. (4.13) into two parts, the temperature and the strain sensitivity, and these two parts can be analyzed separately [1].

4.2.1 TEMPERATURE SENSITIVITY

A shift in Bragg wavelength due to the change in temperature can be done in two ways. First, due to the thermal expansion/contraction, this leads to change the grating period Λ and thus changes the Bragg wavelength [1, 2]. The second is due to the effective refractive index mode n_{eff} of the fiber, which is temperature dependent, and also affects the Bragg wavelength value [1, 2]. The shift in the center wavelength of a fiber Bragg grating due to the temperature change can be written as [1, 2, 7]

$$\lambda_B = 2\left(\Lambda\frac{\partial n_{\text{eff}}}{\partial T}+n_{\text{eff}}\frac{\partial\Lambda}{\partial T}\right)\Delta T \qquad (4.14)$$

By substituting for the $(\partial n_{\text{eff}}/\partial T)=\xi n_{\text{eff}}$, where ξ represents the thermo-optic coefficient [1], and $(\partial\Lambda/\partial T)=\alpha\Lambda$, where α represents the thermal expansion coefficient of the fiber [1], and compensation for the Bragg condition Eq. (4.14) can be rewritten as [1, 2]

$$\frac{\Delta\lambda_B}{\lambda_B}=(\xi+\alpha).\Delta T \qquad (4.15)$$

The typical values of ξ and α for SOF and PMMA POF are $(8.6\sim9.3)\times10^{-6}/^{\circ}C$, $0.5\times10^{-6}/^{\circ}C$ and $(-2.567\sim-1.4)\times10^{-4}/^{\circ}C$, $(5\sim8)\times10^{-5}/^{\circ}C$, respectively [1, 2, 4, 5].

Figure 4.3 (a) and (b) show the effect of temperature variations on the Bragg wavelength for SOF and POF Bragg gratings, respectively [7]. It is clear that the response is linear and there is no hysteresis effect. Due to the negative temperature coefficient, POF has a negative slope of the temperature response [3]. This is in contrast to the temperature response of SOF which has a positive slope [7]. It can be seen also that almost 10 nm tuning range can be achieved when the POF temperature is heated up from 25°C (it is assumed the reference temperature) to 75°C, which is larger than the few nanometers achieved in SOF. Furthermore, in order to increase the tuning range for SOF, it is required to increase the grating temperature by several hundreds of times than that for POFs [7].

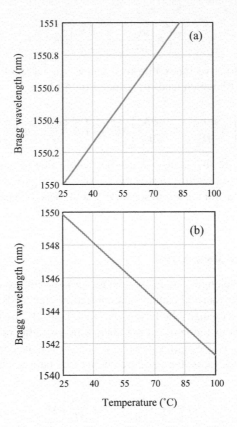

FIGURE 4.3 Temperature effect on Bragg wavelength of (a) silica optical fiber and (b) polymer optical fiber [7].

4.2.2 STRAIN SENSITIVITY

The implications of the applied strain to the value of the Bragg wavelength are two-fold. First, whenever the strains vary, the Bragg wavelength will shift due to expanding or compressing the fiber grating, which leads to change in the grating period [1]. Second, the refractive index of the fiber will be changed, due to the applied strain; this phenomenon is commonly known as the strain-optic effect [7]. The differential shift in center wavelength of a fiber Bragg grating due to the change in the applied strain can be written as [1, 2]

$$\Delta\lambda_B = 2\left(\Lambda\frac{\partial n_{\text{eff}}}{\partial L_g} + n_{\text{eff}}\frac{\partial\Lambda}{\partial L_g}\right)\Delta L_g \tag{4.16}$$

By assuming $\left(\Delta L_g / L_g\right) = \varepsilon_z$ and substituting it in Eq. (4.16), we get

$$\Delta\lambda_B = 2\Lambda\frac{\partial n_{\text{eff}}}{\partial L_g}\Delta L_g + 2n_{\text{eff}}\frac{\partial\Lambda}{\partial L_g}\varepsilon_z L_g \tag{4.17}$$

By using the facts that $\dfrac{\partial n_{\text{eff}}}{\partial L_g}\Delta L_g = \Delta n_{\text{eff}}$ and $\Delta\left(\dfrac{1}{n_{\text{eff}}^2}\right) = -\dfrac{2\Delta n_{\text{eff}}}{n_{\text{eff}}^3}$ [1, 7], we can rewrite Eq. (4.17) as

$$\Delta\lambda_B = 2\Lambda\left[\frac{n_{\text{eff}}^3}{2}.\Delta\left(\frac{1}{n_{\text{eff}}^2}\right)\right] + 2n_{\text{eff}}\frac{\partial\Lambda}{\partial L_g}\varepsilon_z L_g \tag{4.18}$$

The strain-optic effect in an optical fiber results in a change in the effective refractive index given by [1, 7]

$$\Delta\left(\frac{1}{n_{\text{eff}}^2}\right)_i = \sum_{j=1}^{3} p_{ij}.S_j \tag{4.19}$$

In Eq. (4.19), p_{ij} is the strain-optic tensor and S_j is the strain vector [1]. The strain vector S_j for a longitudinal strain along the fiber grating axis (z-axis) is given by [8, 9]

$$S_j = \begin{bmatrix} -v\varepsilon_z \\ -v\varepsilon_z \\ \varepsilon_z \end{bmatrix} \tag{4.20}$$

where v is Poisson's ratio and ε_z represents strain in the z-direction. For a typical germanium-silicate optical fiber, p_{ij} has only two numerical values, p_{11} and p_{12} [1]. Thus, Eq. (4.19) is equal to

$$\Delta\left(\frac{1}{n_{\text{eff}}^2}\right)_i = \left[p_{12} - v\left(p_{11} + p_{12}\right)\right].\varepsilon_z \tag{4.21}$$

Also, since the $\left(\partial\Lambda/\partial L_g\right) = \left(\Lambda/L_g\right)$ [1], then Eq. (4.18) can be rewritten as

$$\Delta\lambda_B = 2n_{\text{eff}}\Lambda\varepsilon_z\left\{1 - \frac{n_{\text{eff}}^2}{2}\left[p_{12} - v\left(p_{11} + p_{12}\right)\right]\right\} \tag{4.22}$$

And therefore, the differential shift in the Bragg wavelength due to the applied strain is given as [1, 2]

$$\frac{\Delta\lambda_B}{\lambda_B} = \left(1 - p_{\text{eff}}\right).\varepsilon_z \tag{4.23}$$

where p_{eff} is the index-weighted effective strain-optic coefficient given by [1, 2, 7]

$$p_{\text{eff}} = \frac{n_{\text{eff}}^2}{2}\left[p_{12} - v\left(p_{11} + p_{12}\right)\right] \tag{4.24}$$

The typical values of P_{11}, P_{12} and v for SOF and PMMA POF are 0.113, 0.252, 0.16, and 0.3, 0.297, 0.35, respectively [1, 2, 7, 10, 11].

Figure 4.4 (a) and (b) show the effect of strain on the Bragg wavelength of SOFs and POFs, respectively [7]. As shown, the Bragg wavelength shift in POF over a

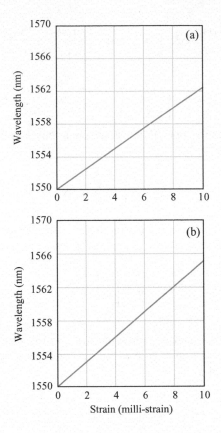

FIGURE 4.4 Effect of strain on Bragg wavelength of (a) silica optical fiber and (b) polymer optical fiber [7].

strain range of 10 milli-strain is more than 15 nm. In addition, the strain sensitivity of the POFs is found to be 1.48 pm/με. This is nearly 1.2 times larger than the value for the SOFs, which is 1.2 pm/με at the designed wavelength [7]. This is because the strain sensitivity of POF is more than of SOF. For example, the Young's modulus of POF is more than 30 times smaller, and its break-down strain is also much larger than for SOF. Thus, the tunability of POF is higher than that of SOF [3, 7].

4.3 TEMPERATURE DEPENDENCE OF THE STRAIN SENSITIVITY

Many researchers have reported on the strain sensitivity of a fiber Bragg grating sensor. However, in all these reports, the strain sensitivity of the Bragg wavelength is generally assumed to be independent of temperature [4, 5, 8, 11, 24]. However, it is known that the material parameters that determine the strain sensitivity, the effective strain-optic coefficients p_{ij} and Poisson's ratio v, are themselves temperature dependent– [7–15]. The dependence of the Bragg wavelength λ_B upon strain ε_z and temperature T changes after taking into account the temperature dependence of the effective strain-optic coefficients p_{ij}, and Poisson's ratio v is given by [7]

$$\lambda(T,\varepsilon_z) = \lambda_B \left\{ 1 + \varepsilon_z - \frac{n_{\text{eff}}^2(T)}{2} \varepsilon_z \left[p_{12}(T) + \left(p_{11}(T) + p_{12}(T) \right) v(T) \right] + \left(\xi + \alpha \right) T \right\}$$

(4.25)

Equation (4.25) represents the modified Bragg wavelength resulting from longitudinal-applied strain and temperature changes. Using Eq. (4.25), the temperature dependence strain sensitivity can be given [7]

$$\frac{\Delta\lambda(T)}{\Delta\varepsilon_z} = \lambda_B \left\{ 1 - \frac{n_{\text{eff}}^2(T)}{2} \left[p_{12}(T) + \left(p_{11}(T) + p_{12}(T) \right) v(T) \right] \right\}$$

(4.26)

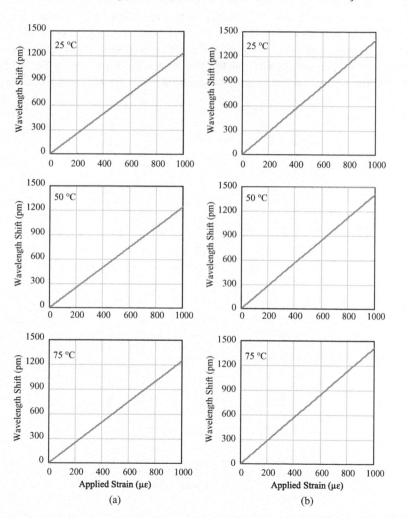

FIGURE 4.5 Temperature effect on strain response of (a) silica optical fiber and (b) polymer optical fiber [7].

According to Eq. (4.26), the temperature dependence of the fiber Bragg grating strain sensitivity arises from the temperature sensitivity of the strain-optic coefficients and of Poisson's ratio [7]. Equation (4.26) was used to investigate the effects of temperature on the fiber Bragg grating strain response [7].

Figure 4.5 (a) and (b) show the dependence of the Bragg wavelength (λ_B) upon strain and temperature variations for the SOF and the POF, respectively [7]. Results show that there is a change in the response of the FBG at different temperatures, though the magnitude of this change is such that its effect would only be significant over large temperature or strain ranges. Also, results indicate that the dependence of the Bragg wavelength (λ_B) upon strain and temperature variations for the POF Bragg gratings lies within the range of 0.14–0.15 fm $\mu\varepsilon^{-1}$ °C^{-1} and in the range of 0.462–0.470 fm $\mu\varepsilon^{-1}$ °C^{-1} for the SOF and the POF, respectively [7].

Figure 4.6 (a) and (b) show the temperature dependence of the strain response for POF Bragg gratings and SOF Bragg gratings, respectively [7]. As we have mentioned, the temperature dependence of the strain response arises from the temperature-strain-optic coefficient and the Poisson's ratio dependence. Results show good linearity over the range of temperature used. Also, results show that the strain

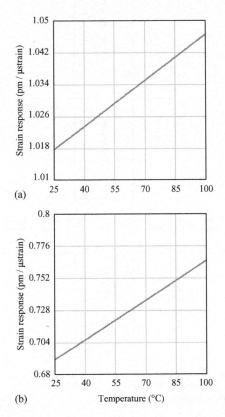

FIGURE 4.6 Effect of temperature variations on strain response for (a) silica optical fiber and (b) polymer optical fiber [7].

response for the POF Bragg gratings changed on average by 1.034 ± 0.02fm $\mu\varepsilon^{-1}°C^{-1}$ and on average by 0.36 ± 0.03fm $\mu\varepsilon^{-1}°C^{-1}$ for the SOF Bragg gratings [7].

4.4 SUMMARY

In this chapter, the mechanisms used to sense fiber Bragg gratings have been discussed after reviewing the modeling simulation methods. The sensitivity of the fiber Bragg grating due to the temperature and strain variation has been investigated by comparison between silica and polymer optical fibers. Also, the temperature dependence of the strain sensitivity has been discussed.

REFERENCES

1. Othonos, A., and Kalli, K. *Fiber Bragg Gratings: Fundamentals and Applications in Telecommunications and Sensing*, Artech House, Boston, MA, 1999.
2. Kashyap, R. *Fiber Bragg Gratings*, Academic Press, New York, NY, 2009.
3. Hill, K.O., and Meltz, G. Fiber Bragg grating technology fundamentals and overview. *J. Lightwave Technol.* 1997, 15, 1263–1276.
4. Lemaire, P.J., Atkins, R.M., Mizrahi, V., and Reed, W.A. High-pressure H2 loading as a technique for achieving ultrahigh UV photosensitivity and thermal sensitivity in GeO2 doped optical fiber. *Electron. Lett.* 1993, 29, 1191–1193.
5. Stone, J. Interaction of hydrogen and deuterium with silica optics, a review. *J. Lightwave Technol.* 1987, 5, 712–733.
6. Xiong, Z., Peng, G.D., Wu, B., and Chu, P.L. Highly tunable Bragg gratings in single-mode polymer optical fibers. *IEEE Photon. Technol. Lett.* 1999, 11, 352–354.
7. Hisham, H.K. Effect of temperature variations on strain response of polymer Bragg grating optical fibers. *Iraq. J. Electr. Electron. Eng.* 2017, 13, 53–58.
8. Eldada, L., and Shacklette, L.W. Advances in polymer integrated optics. *IEEE J. Sel. Top. Quantum Electron.* 2000, 6, 54–68.
9. Liu, H.Y., Peng, G.D., and Chu, P.L. Thermal stability of gratings in PMMA and CYTOP polymer fibers. *Opt. Commun.* 2002, 204, 151–156.
10. Yu, J.M., Tao, X.M., and Tam, H.Y. Trans-4-stilbenemethanol-doped photosensitive polymer fibers and gratings. *Opt. Lett.* 2004, 29, 156–158.
11. Yuan, W., Stefani, A., Bache, M., Jacobsen, T., Rose, B., Herholdt-Rasmussen, N., Nielsen, F.K., Andresen, S., Sørensen, O.B., Hansen, K.S., and Bang, O. Improved thermal and strain performance of annealed polymer optical fiber Bragg gratings. *Opt. Commun.* 2011, 284, 176–182.
12. Kuzyk, M.G. *Polymer Fiber Optics: Materials, Physics, and Applications*, CRC Press, Boca Raton, FL, 2007.
13. Kuriki, K., Koike, Y., and Okamoto, Y. Plastic optical fiber lasers and amplifiers containing lanthanide complexes. *Chem. Rev.* 2002, 102, 2347–2356.
14. Yu, J.M., Tao, X.M., and Tam, H.Y. Fabrication of UV sensitive single-mode polymeric optical fiber. *Opt. Mater.* 2006, 28, 181–188.
15. Erdogan, T. Fiber grating spectra. *J. Lightwave Technol.* 1997, 15, 1277–1294.
16. Measures, R.M. Smart composites structures with embedded sensors. *Compos. Eng.* 1992, 2, 597–618.
17. Kersey, A.D., Davis, M.A., Patrick, H.J., LeBlanc, M., Koo, K.P., Askins, C.G., Putnam, M.A., and Friebele, E.J. Fiber grating sensors. *J. Lightwave Technol.* 1997, 15, 1442–1463.

18. Butter, C.D., and Hocker, G.B. Fiber optics strain Gauge. *Appl. Opt.* 1978, 17, 2867–2869.
19. Scholze, H. *Glass; Nature, Structure, and Properties*, Springer-Verlag, New York, NY, 1991.
20. Barlow, A.J., and Payne, D.N. The stress optic effect in optical fibres. *IEEE J. Quantum Electron.* 1983, 19, 834–839.
21. Sirkis, J.S. Unified approach to phase-strain-temperature models for smart structure interferometric optical fiber sensors: Part 1, Development. *Opt. Eng.* 1993, 32, 752–761.

5 Fiber Optic Sensors for Oil and Gas Applications

5.1 INTRODUCTION

Fiber optic sensor technology has revolutionized the monitoring of wells and reservoirs in the oil and gas industry. This development is due to the passive nature of fiber optic sensors, its ability for cost-effective installations, and the possibility of measurements along the fiber length. The information obtained through fiber optic sensors installed in oil and gas wells effectively contributes to improved efficiency, safety, and final recovery [1].

Over the past five years, the oil and gas industry has seen a rapid increase in the use of fiber optic sensor technology to monitor the bottom of wells. The significant development in the glass chemistry industry enabled filling up to 20 fibers in a very narrow cable slot which contributed to the promotion of the analysis applications for strong bottom well sensing. Each of the fibers can be turned into a fully distributed sensor for temperature (DTS), acoustics (DAS), and strain (DSS) measurements, or these can be used to interrogate multiple point sensors like pressure gauges and geophones [1].

Fibers can also be used for downhole power transmission and data that is measured remotely. Although conventional devices have been used for decades to gather information, the new technology offers many advantages. The most important is that the operator can now observe the entire well simultaneously and can see and hear objects in the well hole that were impossible by using traditional systems [1]. Another is that glass fiber by its nature is immune to electromagnetic radiation, can transmit data at high speeds, and can operate in extreme environments. All of these attributes equal a very attractive package that can benefit most well-monitoring scenarios [1, 2].

The information collected is equal to the amount that the decision maker wishes to pay for it before making a critical decision. As a result of the continuous increase in unconventional reservoirs, the need for rapid subsurface learning has become indispensable, with much of the decades of learning from conventional reservoirs not applicable. In contrast, unconventional reservoirs are being rapidly drilled and developed in both delineation and factory mode, with the windows of opportunity to make asset optimization decisions very short [1].

During the monitoring process, operators attempt to improve and optimize the data and track spacing adjustment, phase combinations as well as fluid chemistry, selection of filler support, and many other variables. Low recovery factors and high costs of completion and motivation support the need to determine the optimal combination of variables early in the project. Regardless of how many models are generated to estimate what will happen, even these models are calibrated, with field data becoming truly valuable [1].

The use of fiber optic sensor technology has answered many important questions without having to enter the well hole, making it appear as a very attractive option for the oil and gas industry. Compared to conventional methods, the use of optical fiber fittings in reservoirs is not limited to one event or phase, but most of the life cycle of the well is covered entirely. In fact, most of the life cycle of a well is covered, from monitoring cement to monitoring completion to controlling stimulation to evaluating production and finally for identifying and monitoring re-stimulation candidates. The subsurface insight gathered by fiber optics during these different phases of a well is allowing operators to make quicker asset optimization decisions in shale plays [1].

5.2 DOWNHOLE MONITORING

Compared with traditional vertical wells drilled for use in oil industries, modern wells used to extract hydrocarbon reserves in fields with high production rates usually have complex configurations including vertical, oblique, multi-lateral/lateral branches, or components [2]. In addition, the tank can be divided into several levels, each producing different rates of gas, oil, water, pressure, and temperature. The engineer's vision of the tank shows his ability to extract most of the oil components before reducing the reservoir to a water product, and abandon it. The problem is that the average recovery efficiency is almost 35%, meaning that 65% of the oil or gas is left in the tank [2].

To improve the recovery efficiency of the oil wells, the industry must move toward smart innovations to optimize its performance. To achieve this, additional sensory capacity must be added to the environment of the borehole on a permanent basis [2]. On the other hand, the unique characteristics of fiber optic sensors are ideal for monitoring the wells of multi-fiber or distributed wells, especially for applications at the bottom of the well. Is there a need to monitor parameter or parameters in many spatial locations through the well hole or horizontal/of the well of interest [3].

5.2.1 RESERVOIR PRESSURE AND TEMPERATURE MONITORING

Pressure and temperature are the main parameters of reservoir engineering, where the permanent monitoring of the downhole is widely used. Although traditional pressure and heat sensors are used to manage reservoirs widely, many limitations limit their performance. Essentially, many of these limitations are due to the electronics specifications required at the bottom of the well, where conventional pressure and heat sensor technology is unreliable at high temperatures [3].

Also, the limitations of multiplexing limit the spatial accuracy of conventional sensors. In contrast, fiber optic sensors have high reliability and accuracy at high operating temperatures, making them very suitable for working at the downhole and conducting measurements at high temperatures [3].

Kersey [2] illustrates an example of the calibration properties of the FBG-based pressure sensor and the developer to monitor production using mechanical translation to compress to stress in fibers. Kersey shows the CiDRA pressure packaged [2]. The sensors show highly linear response in wavelength shift with pressure. These sensors have been tested for temperatures up to approximately 175°C and work is ongoing to develop them and enable them to operate up to 250°C as the design target [2].

In Kersey [2], the high accuracy achieved with a 5,000 psi-range FBG-based sensor have been shown. Where the residual error due to the pressure and temperature measuring is presented. The error from calibration is within approximately +/−1 psi over the full operating range. This accuracy level is comparable to the performance found with the best electrical gauges used in the industry [2, 3].

5.2.2 FLOW MONITORING

Multi-stage flow meters measure the flow rates of the individual components of the flowing mixture of oil, gas, and water without having to separate the components as shown in [3, 4]. Over the last few years, the industry has been mostly focused on multi-stage flow meters and subsea layers, resulting in the technology being made so clear that these counters are commercially available. Then the industry witnessed an unprecedented development, where the flow meters h been made moveable downhole [4, 5]. Their ability to maintain continuous monitoring of the well production makes the bottom well flow meters provide additional potential to determine flow rates based on spatial distribution from inside the well. However, the challenges associated with the development of multi-path flow meters are accurate and reliable at the downhole [4]. Several harsh operating conditions, multi-phase flow systems, restricted packing constraints, data transfer challenges, environmental constraints, and non-parasitic requirements have proven to be very difficult barriers to overcome [3].

Fiber optic sensing technology provides the possibility of conducting these measurements at the bottom of the well [5]. The ability to exploit the potential of fiber optic sensors to accommodate extreme environmental conditions such as temperature and high pressure and data transmission requirements; multi-phase fiber optic flow meters have been developed for this purpose. The flow meter is non-volatile, intrinsically safe, and does not contain electronic determinants that restrict work at the bottom of the well [5]. In the long run, the goal is to develop multi-phase flow meters capable of accurate measurement across an expanded range of multi-phase flow systems [4]

5.2.3 SEISMIC MONITORING

Fiber optic sensors have huge potential to provide distributed sensors for acoustic pressure in green environments for seismic observation [6]. The current trend in the industry is taking advantage of the surface imaging capabilities of seismic monitoring, not only for oil and gas exploration but also to monitor the estimated life of the depletion of the reservoirs. This 4D seismic interval is a powerful new approach to increasing production in the oil and gas industry [6]. Installed devices permanently located deep in wells can provide such imaging, but the challenge is their ability to survive for more than 10 years. Again, their unique properties help them, as the negative nature of fiber optic sensors allows high reliability in harsh environments over long periods of time [6].

When traditional methods of measuring the circulation in the seismic field are used, the results are most likely to be susceptible to disturbances due to the high sensitivity of these instruments to linear movements [7–9]. The results, therefore, do not have the high precision that can be relied upon, which necessitates the search for promising new tools with high reliability for measuring the rotational components of earth motion [10].

In practice, many known techniques are used to detect changes in circulation, but those that rely on the principle of Sagnac effect, which lack inertia and which have a major advantage, detect the rotation itself. Two types of gyroscope systems can be distinguished: ring lasers [11] and optical fiber seismic scales [12, 13], both of which depend on the technical implementation of the Sagnac interference [14].

The seismic sensor based on fiber optic technology dynamically measures the strain of the fiber between two FBGs using the combined configuration technique of an interferometric technique and a time domain multiplexing (TDM) to transmit the dynamic fiber strain information to the recording instruments [14]. FBG is used to separate fiber sections into individual sensors which allow multi-sensors on a single fiber for recording and analysis. The low FBG reflectivity allows the light to continue through the fiber to the next set of FBGs, and thus many sensors [14].

This combination, based on fiber optic technology, provides a large number of seismic sensors by diffusing on a single fiber with a churn on the high performance characteristics of the sensor [14]. Because it is immune to electrical and electromagnetic interference, the system does not require any electronics at the end of the fiber optic sensor. In addition, this design of fiber optic sensors makes them extremely powerful and able to operate in harsh environments such as very high temperatures that may reach more than 300°C [14]. Even in higher temperature conditions it can operate using specialized fibers. The TDM method probes the sensors by sending one optical pulse at a time and recording the reflections of FBG from each sensor in a matrix [14].

The principle of seismic sensor operation is based on the measurement of strain in the interference ally by comparing the changes in the relative phase angle between the FBG reflections in a portion of the sensor fibers. In the case of FOSs, the fiber sensed responds to seismic vibration through dynamic fiber stress [14]. The fiber optic sensor system can measure strains in fibers at very high resolution that may reach less than 1 Angstrom ($1 \times 10\text{-}10$ m). The principle of the telemetry system is based on low-noise sensing by nature because it does not capture electrical noise from any source. The system also uses low-noise electronics to convert optical data to electrical digital data [14].

5.2.4 PIPELINE MONITORING

Pipelines represent the backbone of the lifestyle of modern and irreplaceable societies, as they serve water, gas, oil, and all kinds of products [15, 16]. Therefore, failures in these systems not only result in service interruption and financial losses but also cause spills that cause environmental pollution or even catastrophic events [16]. These reasons have made governments, engineering companies, and industry associations work hard to develop the standards and designs for pipelines and operate them with rules so that leaks can be significantly reduced since the 1960s and early 1970s. As a result, pipelines today are the most reliable and safe mode of transport [16].

Typically, fiber optic cables are designed so that scattering effects are reduced to as little as possible in order to increase the transmission distance and the rate of transmitted data. Some of the dispersion effects of injected laser light depend on the conditions surrounding the fiber optic cable, such as temperature and strain [16].

The scattering effects of Raman and Brillouin can be obtained by using advanced and specialized optical time-domain reflectometers (OTDR). Short laser pulses are sent from the measurement device into the fiber for analyzing to determine the time–distance related reflection/scattering signals with regards to frequency and amplitude of the desired scattering effect [17]. As a result, strain and temperature measurement becomes possible along the fiber [17].

5.2.5 POWER CABLE AND TRANSFORMER MONITORING

In practice, the insulation capacity of power cables is usually classified at a 90°C operating temperature. Particularly in power cable tunnels, where cables and fittings are assembled into shelves, temperatures in these cases can exceed specified values [18]. For this reason, temperature sensors are installed with optical fibers inside the cable insulator so that they are able to conduct a high-temperature direct assessment, so countermeasures can be initiated easily [18].

5.2.6 STATUS MONITORING OF WATER MAINS

Pre-stressed concrete cylinder pipes (PCCPs) are widely used in water pipes. In many cases, however, they did not appear to be efficient, as many major water ruptures occurred, with significant damage and a large number of faults resulting in the loss of large amounts of water [18,19]. The PCCP stability is mainly related to the number of broken wires inside the tube. When a particular wire is cut off, it will emit a special sound, which can be detected by a sensitive optical sensor installed inside the pipeline [18].

Depending on the tube book, a special software package can calculate the number of broken wires so that they can issue a special warning if they exceed a certain threshold or the number of wires broken over a given time period to a maximum limit [19]. The big challenge is that the wire-cutting period is so short that conventional methods, such as acoustic detectors, fail to obtain sufficient information to clearly identify a wire break event [19]. In contrast, other interferometers such as Sagnac or Michelson interferometer continuously analyze the signals by receiving the information available about the sound of the wire break; however, because the signal is weak, it fails to locate the signal. An interferometer and a built-in Brillouin amplifier were then described and applied to several major water pipelines in the United States of America [19].

5.3 FIBER OPTICS SENSORS FOR DETECTION APPLICATIONS

Based on the above, compared to the traditional methods used to control pipes and shelters, referred to as the speed of fracture or cracking and failure to determine it, fiber optic sensors are an indispensable option for pipeline control in many applications. In practice, many of these applications have been implemented in recent years throughout the industry [19].

5.3.1 LEAK DETECTION

Typically, the loss of the medium transferred by leakage in the oil pipelines results in one or more of the following effects that can be detected: local cooling due to the Joule–Thomson effect, [19] and soil temperature change due to differences in temperature between the soil and the liquids emitted as a result of leakage . Because of evaporation effects, especially in high-pressure applications, the resulting medium produces detectable sounds [20].

Temperature changes can be detected by dependence on the effects of Raman or Brillouin dispersion [17] on ways to compare average temperature with soil temperature [17, 18]. The temperature of natural gas, saline solution, phenol, sulfur, liquefied natural gas, crude oil, and other media is then reported by means of sensors which allow the detection of very small leaks [20, 21].

Compared to conventional pipe control methods, this technology has an added advantage that makes it reliable and completely independent of any operating constraints. Even during the periodic opening and closing of small leaks in gas pipelines due to the effects of freezing, they can be identified using modern signal analysis methods [20]

5.3.2 GROUND MOVEMENT DETECTION

The movement of the ground layers causes changes in nature such as geological storms, earthquakes, and landslides. These factors lead to increased pressure on pipelines, tunnels, and other underground infrastructure [21]. To avoid this legacy, fiber optic sensors were used in two ways to identify threatened terrestrial movements [22]:

- Optical fibers are directly linked to the strain directly from the pipeline walls to measure changes in the wall strain and to identify the resulting movements and distortions. If a sudden increase in the strain is detected, it can be treated by reducing the internal pressure of the tubes and thus reducing the overall stress in order to reduce the risk and/or the effects of leakage [22–24].
- Fiber optic sensor cables are installed in parallel and close to the infrastructure to be monitored, allowing large road sections to be covered by simple and efficient cable installation [25].

As a result of increasing ground movements, it has been observed that pipeline sections have increased risk by point sensors in recent years and can therefore be monitored by installing fiber optic sensors distributed along the longer pipeline extensions [25]. Therefore, changes in stress due to ground work can be monitored as planned, such as the concrete installation of pipes and cross cables [25].

5.3.3 Fire Detection

The temperature sensor can be used with fiber optic cables as a thermal detector to detect sudden rise in temperature or fire in tunnels [26]. Since the optical cable is sensitive in its full length and since the temperature can be detected within a wide range, it is possible to determine the location of the fire in detail. This leads to the treatment of high temperature through ventilation so that it can be controlled and coordinated efficiently [26].

5.3.4 Pig Position Detection

The designated systems which are designed to detect sounds of third party activities are also apt to detect the sounds created by pigs according to Liu and Kim [27].

5.4 GAS NETWORK MONITORING

For the gas distribution networks, the challenges of cracks and leaks are difficult problems and must be resolved urgently [28]. Physical signals such as temperature change, leakage noise, etc. caused by low-pressure gas leakage are weak and difficult to detect by conventional methods. According to a study [28], the cooling effect of Joule–Thomson during the gas leakage process leads to a drop in the gas temperature [29]. This temperature drop can be detected by DTS in the vicinity of dropout points. Potential anomalies can therefore be identified. In this section, calculations are made to assess changes in the temperature of the leachate gas caused by potential leaks in different sections of the underground gas network [30]. Note that argon gas contains a Joule–Thomson coefficient similar to methane. By taking safety considerations, argon gas is selected as a gas for calculation to facilitate possible experimental verification [31–33]. The assessment procedure for the gas temperature leak is shown in [29].

If there is sufficient source gas, the Joule–Thomson effect will occur continuously and automatically. Thus, during the gas leak, its flow will affect the surrounding ground and will spread more into porous soil, as shown in [34]. Prolonged leakage of gas can also reduce the air temperature in the area near the cracks during heat transfer [34].

Last, the ambient temperature in the vicinity of the cracks can be close to the low temperature of the gas leak resulting from the impact of Joule–Thomson, resulting in the possibility of detecting the change in ambient temperature by the distributed temperature sensor installed along the tube. The warning signals are then sent quickly to the control room of the engineer's flags to avoid potential risks and to do the necessary and to avoid further loss of property [34]. The gas leak can be better detected by improving the DTS warning system [30, 34].

One important point to note is that response time is not regarded as the first important parameter to be relied upon, since there is a certain amount of leakage that is acceptable and the focus should be on precise settlement of the gas leakage points [34]. The rapid response time may weaken the temperature accuracy, which is the critical parameter for detecting slight temperature changes [34]. The main point of the DTS warning system should also be to identify a slightly low temperature signal for a long period of time in order to avoid unnecessary interference of noise signals that may send erroneous alarms [35].

5.5 SUMMARY

In this chapter, the application of fiber Bragg gratings for oil and gas sensing technology was reviewed. This technology includes monitoring (i.e. downhole reservoir pressure and temperature monitoring, oil and gas flow rate monitoring, seismic monitoring, pipeline monitoring, power cable and transformer monitoring, status monitoring of water mains and gas network monitoring) and detection (i.e. leak detection, ground movement detection, fire detection, pig position detection).

REFERENCES

1. Ranjan, P. and McColpin, G. Fiber optic sensing: Turning the lights on downhole. *The Digital Oil Field*, 2013.
2. Kersey, A.D. Optical fiber sensors for permanent downwell monitoring applications in the oil and gas industry. *IEICE Transactions on Electronics* 2000, E83–C (3).
3. John, A. and Igimoh, J. The design of wireless sensor network for real time remote monitoring of oil & gas flow rate metering infrastructure. *Int. J. Sci. Res. (IJSR)*, 2015, 5 (10), 1–14.
4. Aspelund, A. et al. Challenges in downhole mulitphase flow measurments. In: SPE 35559, Presented at the European Production Operations Conference and Expositon, Stavanger, Norway, 1996.
5. Brill, J.P. Multiphase flow in wells. *J. Pet. Technol.* 1987, 39 (1), 15–21.
6. Kurzych, A., Jaroszewicz, L.R., Krajewski, Z., Teisseyre, K.P., and Kowalski, J.K. Fibre optic system for monitoring rotational seismic phenomena. *Sensors* 2014, 14 (3), 5459–5469.
7. Teisseyre, R., and Nagahama, H. Micro-inertia continuum: Rotations and semi-waves. *Acta Geophys. Pol.* 1999, 47, 259–272.
8. Jaroszewicz, L.R., Krajewski, Z., Solarz, L., Marc, P., and Kostrzyński, T. A new area of the fiber-optic Sagnac interferometer application. In: Proceedings of the 2003 SBMO/IEEE MTT-S International Conference on Microwave and Optoelectronics, Rio de Janeiro, Brazil, pp. 661–666, 20–23 September 2003.
9. Sagnac, G. The light ether demonstrated by the effect of the relative wind in ether into a uniform rotation interferometer. *Acad. Sci.* 1913, 95, 708–710. (in French).
10. Schreiber, U., Schneider, M., Rowe, C.H., Stedmanand, G.E., and Schluter, W. Aspects of ring lasers as local earth rotation sensors. *Surv. Geophys.* 2001, 22, 603–611. Sensors 2014, 14 5469.
11. Jaroszewicz, L.R., Krajewski, Z., and Solarz, L., Absolute rotation measurement based on the Sagnac effect. In: *Earthquake Source Asymmetry, Structural Media and Rotation Effects*, Vol. 31, Teisseyre, R., Takeo, M., Majewski, E., Eds. Springer: Berlin, Germany, pp. 413–438, 2006.
12. Takeo, M., Ueda, H., and Matzuzawa, T. Development of a High-Gain Rotational-Motion Seismograph. Grant 11354004. Earthquake Research Institute University of Tokyo: Tokyo, Japan, pp. 5–29, 2002.

13. Nigbor, R.L., Evans, J.R., and Hutt, C.R. Laboratory and field testing of commercial rotational seismometers. *Bull. Seismol. Soc. Am.* 2009, 99 (2B), 1215–1227.
14. Spillman, W.B., Huston, D.R., and Wu, J. Fiber optic sensors for seismic monitoring. In: Teisseyre, R., Takeo, M., Majewski, E., Eds. *Earthquake Source Asymmetry, Structural Media & Rotation Effects*, Springer: Berlin, Germany, pp. 521–545, 2006.
15. Paulsson, B.N.P., Toko, J.L., Thornburg, J.A., Slopko, F., He, R., and Zhang, C.-h., A high performance fiber optic seismic sensor system. In: Proceedings, Thirty-Eighth Workshop on Geothermal Reservoir Engineering Stanford University, Stanford, California, February 11–13, 2013.
16. Frings, J. Enhanced pipeline monitoring with fiber optic sensors. In: 6th Pipeline Technology Conference, 2011.
17. Inaudi, D., Glisic, B., Figini, A., and Walder, R. Pipeline leakage detection and localization using distributed fiber optic sensing. In: RIO Pipeline Conference, 2nd to 4th October 2007.
18. Waldner, R. Pipeline leakage detection and localization using distributed fiber optic sensing. Webinar. s.l.: Smartec, 2009.
19. Elliot, J., Stieb, J., and Holley, M. An integrated dynamic approach to PCCP integrity management. Pipelines - Service to the Owner, 2006.
20. Paulson, P.O. Fiber Optic Sensor Method and Apparatus. US Patent 7,564,540, B2. United States, 21 July 2009.
21. Sensornet. *Digital Pipeline Integrity Monitoring System - Application Guide.*
22. Schlumberger, 2008. *Integrated Pipeline Monitoring – Integrity.*
23. Long-range pipeline monitoring by distributed fiber optic sensing. D. Inaudi, B. Glisic. 6th International Pipeline Conference. s.n.: Calgary, Canada, 2006.
24. Earthquake detection and safety system for oil pipeline. L. Griesser, M. Wieland, R. Walder, *Pipeline & Gas Journal*, December 2004.
25. Daniele, I., and Branko, G. Overview of Fibre Optic Sensing Applications to Structural Health Monitoring. 13th FIG Symposium on Deformation Measurement and Analysis. s.n.: Lisboa, 2008.
26. Omnisens. *Pipeline Integrity Monitoring - Transandean Route, Case Study.*
27. Liu, Z., and Kim, A.K. Review of recent development in Fire detection technologies. *J. Fire Prot. Eng.* 2003, 13 (2), 129–151, National Research Council Canada.
28. Optasense, Pipeline and Security Monitoring. s.l. www.quinetiq.com, 2010.
29. Ai, G., Ng, H.W., and Liu, Y. Study on heat transfer process during leaks of high pressure argon through a realistic crack. *Int. J. Therm. Sci.* 2016, 99, 213–227.
30. Roebuck, J.R., and Osterberg, H. The Joule-Thomson effect in argon. *Phys. Rev.* 1934, 46 (9), 785–790.
31. Burnett, E.S. Experimental study of the Joule-Thomson effect in carbon dioxide. *Phys. Rev.* 1923, 22 (6), 590–616.
32. Roebuck, J.R., and Osterberg, H. The Joule-Thomson effect in helium. *Phys. Rev.* 1933, 43 (1), 60–69.
33. Bridgeman, O.C. The Joule-Thomson effect and heat capacity at constant pressure for air. *Phys. Rev.* 1929, 34 (3), 527–533.
34. Campanella, C.E., Ai, G., and Ukil, A. Distributed fiber optics techniques for gas network monitoring. *Oil & Gas J.* 2010, 108 (1–6).
35. Lapique, F., Meakin, P., Feder, J., and Jossang, T. Relationship between microstructure, fracture-surface morphology, and mechanical properties in ethylene and propylene polymers and copolymers. *J. Appl. Polym. Sci.* 2000, 77 (11), 2370–2382.

6 Fiber Optic Sensors for Civil Applications

6.1 INTRODUCTION

The significant development of fiber optic sensing technology has made it an important option to monitor the performance of civil infrastructure [1]. Their unique advantages, such as small dimensions, high resolution, excellent signal transmission capability over long distances, immunity to electromagnetic and radio frequency interference and ability to incorporate a series of interrogated sensors multiplexed along a single fiber, make it an important choice for civil engineers and a strong competitor than traditional sensors [1, 2]. In this chapter, we will present a review of the most recent developments in the technology of fiber optic sensors as well as some of their applications in monitoring the performance of civil infrastructure such as buildings, bridges, sidewalks, dams, tunnels and piles, etc. [4, 5]. Further, the most prominent current problems faced by fiber optic sensors will be discussed with their applications to monitor civil structural performance [3–5].

6.2 TYPICAL FIBER OPTIC SENSORS IN CIVIL FIELDS

6.2.1 CRACK SENSORS

The dramatic beginning of the failure of concrete structures usually begins with cracks, where, by observing the cracks in the structure, the condition of damaged concrete structures can be assessed [5]. Many techniques have been developed, such as optical inspection, radiography, ultrasound, and sound emissions for damage detection; however, they all share a common limitation: their inability to continuously evaluate cracks in the site during the service of structures [5].

The fiber optic crack sensors (FOCSs) developed have provided a typical solution to this problem, which has been used to detect cracks by a number of researchers. Wanser and Voss [6] used multi-mode optical time-domain reflectometer (OTDR) to measure the 2D crack growth and crack displacement such as longitudinal crack separation and transverse shear crack displacement, respectively. Habel [7] performed a new approach for crack detection based on measuring the attenuation of light transmitted in the FOCSs due to the surface crack growth. Liu and Yang [8] have used the distributed FOCSs for monitoring the concrete cracks based on the light loss due to the micro bending of optical fiber bridging cracks with the use of OTDR.

Lee et al. [9] demonstrated the capability of intensity-based optical fiber sensors (IOFSs) to monitor the fatigue crack growth of steel structures by detecting the stiffness changes near the crack. Although fiber optic crack sensors have been successfully applied in many cases, they suffer from some limitations [2]. Integrated

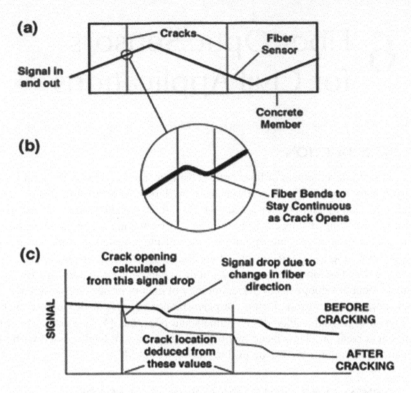

FIGURE 6.1   The novel crack sensing concept [11].

sensors, which measure the displacement between two points separated by a relatively large distance, cannot distinguish the case of many fine cracks and the case of one widely open crack [10].

To overcome these problems, Leung et al. [11] developed a new fiber optic "distributor" that has the ability to detect the formation of cracks without the need to know in advance its precise location, and also the ability to conduct continuous monitoring once formed. This sensor is highly efficient and has the ability to detect and monitor a large number of cracks using very few fibers [11]. Figure 6.1 illustrates the principle of the work of this sensor, as an example of control cracks before forming on the surface of the bridge [11].

6.2.2   STRAIN SENSORS

Due to its small size [11], the fiber optic sensor is either compactly or superficially anchored to various materials, such as concrete, reinforcing steel, steel sheets, fiber reinforced polymer (FRP) strips, and others. In recent years, there have been many studies on the application of fiber Bragg gratings (FBGs) and FP as sensors to monitor structural performance [12]. Many of these studies are based on the ability of fiber optic sensors to measure the internal strain of structures. Grossman and Huang [13] used FP sensors to measure multidimensional stress. Bonfiglioli and Pascale

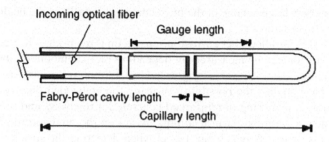

FIGURE 6.2 Scheme of the FP sensor [11].

[14] conducted experiments to use fiber optic sensors on concrete samples for internal measurements [11]. Their results showed a high ability to measure the internal stress of the sample without affecting the overall stress situation due to the small dimensions of the fiber optic sensors.

Also, Kenil et al. [15] have used multiplexed FBG sensors to measure the strains along the 10 mm diameter bars embedded in the reinforced concrete beams subjected to bending. The results obtained showed the high ability of the sensors to measure large strains and strain gradients accurately, without significantly affecting bond properties [15]. In most cases, the physical state of the concrete structure depends heavily on the condition of the strain in the reinforcement bars. Several studies have been conducted on measuring strains on iron bars [16]. Figure 6.2 shows schematic diagram of a FP Sensor. The jacket of the fiber is removed only in the sensor area, and it is attached to the brushed surface of the reinforcing arm by cyanoacrylate. The sensor part is protected by several layers of rubber, and input/output (I/O) lead is protected by fiber jackets (Figure 6.3) [16].

6.2.3 CORROSION SENSORS

The corrosion of steel cables and reinforcing steel in concrete represents one of the leading causes of durability problems affecting civil infrastructures [16]. As a result of the corrosion of reinforcing steel, a large radial pressure is exerted on the surrounding concrete, which may result in local radial cracks. These cracks in turn accelerate the corrosion process of the reinforcement. The detection of corrosion of

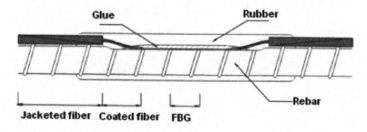

FIGURE 6.3 Scheme of the fiber Bragg grating strain sensor [11].

reinforcement bars has been one of the most challenging tasks in the health monitoring of civil infrastructures [17].

Fuhr et al. [17] installed fiber corrosion sensors on three bridges in Vermont. The degree of corrosion can be measured by depending on the amount of absorption of the diffuse light of the wave that is evaporated by the reinforcing bar. Also, Fuhr and Huston [18] have studied the possibility of using the embedded fiber optic sensors for the corrosion monitoring of reinforced concrete for roadways and bridges. They proposed a new a warning alarm technique based on the predetermined threshold of "fiber events or faults" which can be set when detecting the structure's internal damage (Figures 6.4 and 6.5).

Casas and Frangopol [19] proposed a method for detecting corrosion in a non-corrugated steel rod by using an FBG sensor around a perpendicular circle with superglue. Using this method, the sensor measures the amount of angular pressure produced around the rod. When the tape expands due to corrosion, the perimeter of the section will increase to the FBG sensor. As a result of this change, erosion is detected as a conversion in the waveform of the Bragg sensor [19].

Fiber optics sensors are actually used in concrete. The principle of its work is based on the FP formation to monitor the tensile strain caused by corrosion in the orthogonal concrete with the level of reinforcement bars, where cracks can occur [19].

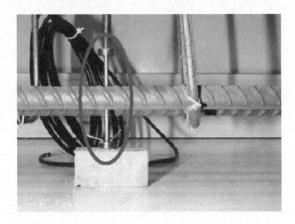

FIGURE 6.4   Location of fiber optic strain sensor [11].

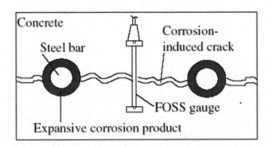

FIGURE 6.5   The concept of using FP sensor to detect corrosion-induced damage [11].

6.3 CIVIL ENGINEERING APPLICATIONS OF THE FIBER OPTIC SENSORS

6.3.1 HIGHWAY STRUCTURES

Bridges and highways are some of the most important infrastructure facilities, and in the event of any faults, they may lead to traffic congestion, resulting in material losses and enormous economic and social costs. Therefore, it is very important to maintain them, and this requires the practice of monitoring continuously to ensure their safety and prevent exposure to any defect. Already, many fiber optic sensors have been installed on bridges and highways around the world [3, 5]. These sensors monitor the mechanical efficiency of the bridges by relying on standards such as temperature, degree of slope, potential for cracks, and others [1–5]. The sensors are positioned point-by-point on important bridge components such as bridge cables, anchors, concrete floors, and sidewalks [2, 3]. Integrated sensors are also installed to measure the curvature of the bridge and its deviation [4]. Distributed sensors are also deployed to monitor the stress level of high-speed vehicles [1–5].

6.3.1.1 Bridges

In practice, when a bridge designed many years ago cannot meet the current need for traffic density, the most economical solution is to rehabilitate the existing structure instead of building a new structure. For example, the bridges north and south of Versailles in Switzerland were typical examples of re-upgrading the bridge [3]. They represent two parallel twin bridges, each supporting two routes of the Swiss National Highway A9 between the cities of Geneva and Lausanne. Their old design is a traditional concrete beam structure in parallel to the pre-stressed supporting concrete surface and two seats. In order to support the absorption of the traffic momentum, they were updated by a third traffic crew and a new emergency route, which led to the expansion of external beams and the spread of the joints [1].

The stages of construction were divided into two parts: internal and external operations, including extension of overhang. The first phase began with the demolition of the internal mattress, followed by one larger reconstruction [1]. The second phase began with the process of destroying the old outer surface and expanding the external network structure and rebuilding it further supported by metal beams [1]. During these processes, more than 100 SOFO interference meters were installed from the long scale on the Versoix bridges to control possible deformation due to the differential shrinkage between old and new concrete (Figure 6.6) [1].

Observations have shown the possibility of knowing the vertical deviation of the beam by using the distortion of its longitudinal trend [1]. This process was done by using a couple of SOFO measurements installed on both the pressure and tensile sides of the same site [1]. The local mean of the curve can be calculated, from which the vertical deviation of the beam can be estimated through integration, on the basis of the appropriate boundary conditions (Figure 6.7) [1].

Fiber grating optic sensors have been used for monitoring the internal strain of the girder box of a pre-stressed concrete bridge across the Ring Canal in Ghent, Belgium, and the trough girder of a railway bridge along the rail track Gent-Moeskroen,

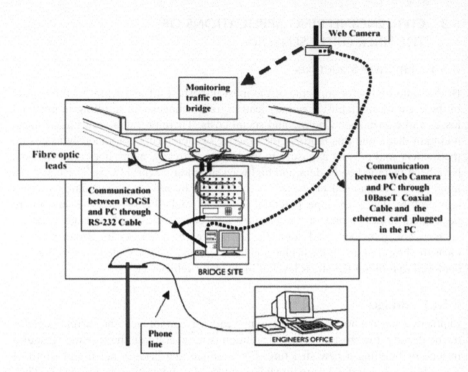

FIGURE 6.6 Monitoring technology for the Taylor Bridge [11].

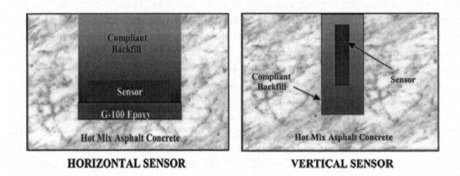

HORIZONTAL SENSOR   VERTICAL SENSOR

FIGURE 6.7 Fiber optic traffic sensors (FOTS) placement options [11].

Belgium, during post-tensioning, proof load tests, and in service load [1]. However, due to the fragility of the sensor, it is impossible to include it directly in the concrete structure. For this reason, different types of packaging techniques have been developed, such as FBG sensors attached to a steel reinforcement rod inside the structure [19]. This sensor consists of a sensor bar of the FBG strain with a thermal amplifier to compensate the temperature [1]. This sensor was attached to a narrow groove on the rod, with adhesive in the middle of the tape, while the thermal bond was connected to the opposite side [1].

The tested results showed that the strain measured by the sensor rods corresponded to the results measured by the demountable mechanical strain gauge (DEMEC) [1]. In addition, there is another way in which the strain and temperature sensors of FBGs can be packaged: by including sensors in FRP [2]. Through this method, stress sensors can measure both tensile stress and compression with high signal-to-noise ratio. So they can measure a strain of up to 5,000 accurate strains with the accuracy of 1–2 microbial strains [2].

6.3.1.2 Dams

The first application of the Brillouin-based distributed optical fiber sensors (DOFS) system for dam monitoring was done by Évenaz et al on the Luzzone dam in the Swiss Alps [16]. The process included setting a temperature distribution map for the dam structure to monitor the amount of time it takes to cool the central area until it reaches 50°C. Regarding the application of Brillouin optical time–domain analysis (BOTDA) and DiTeSt, Inaudi & Glisic reported the Plavinu plug in Latvia [17], where Raman DOFS is more sensitive to temperature change than Brillouin.

These sensors have been applied practically in many dams of Birecik in Turkey; the Shimenzhi arc bridge in Xinjiang, China; the Loyalty Dam and the Mujib Dam in Jordan [18]. These sensors were also used to detect leakage in channels and barriers in Casas et al. [19]. In all applications where this type of sensor was used, it was possible to obtain detailed temperature distributions due to DOFS.

6.3.2 BUILDINGS

Fiber optic sensors have been successfully used to measure many parameters such as stress, temperature, displacement, and cracks in buildings [17–19]. Fuhr et al. [17], have reported that an embedded fiber optic sensor was used for monitoring a 65,000 square foot concrete structure, the Stafford Medical Building at University of Vermont, Burlington. Pressure sensors and wind sensors were installed on the outside of the walls, other sensors were installed on the floor of the building, and support poles were made by connecting them to the wiring using tape [18]. This network of sensors is measured by vibration, wind pressure, loading, crawling, and other parameters related to building performance, such as the development of cracks [18].

In Iwaki et al. [20], a series of photovoltaic sensors was developed to build 12 frames of steel with building techniques to withstand damage. In this model, 64 light sensor units were installed. Using its multicast capability, six sensors were installed on average in one fiber. Displacement, stress, and building temperature were measured [20]. The results obtained are consistent with other studies, in terms of which they support the ideal performance of this type of sensor to monitor the performance of large structures [17–20]. In addition to their application in large buildings, the performance of these sensors in the control of historic buildings is subject to intensive research interests [1].

Four fiber optic sensors were placed on the main arch of Como Cathedral in northern Italy. This building has historical significance as an important cultural heritage; it was built in 1396, and it is therefore important to monitor it for any major structural deterioration [21]. The four sensors were placed across, above, and under the base arch of the building, where each sensor has two connecting networks [21].

Also in Italy, another important historical building dating back to 1600 has been repaired and updated using Brillouin integrated sensors, which are fiber optic sensors that measure the strains by using the Brillouin frequency change [22]. The use of these types of sensors led to a lower cost of monitoring and thus enabled the monitoring of all critical areas at reasonable prices. In addition, the distributed sensor feature allowed detection of anomalies in the transport of the load between the FRP and the substrate, and thus the location of the cracking patterns in the end [22].

6.3.3 GEOTECHNICAL STRUCTURES

Another area that has seen increasing interest in the use of fiber optic sensors is the observation of geotechnical structures such as slopes [1]. The standard used to monitor this type of structure depends on the use of separate sensors, making it difficult to obtain universal slope behavior. In addition, these sensors are often not compatible with the mass deformation of the rocky soil [2]. This type of sensor was used and deployed on the Nanjing Gulu tunnel [23] and the Xuanohuhu Lake tunnel [24], in the Royal China tunnel and London Royal Mail tunnel [25]. Wu et al. also used this type of sensor to monitor the deformation of soil layers and water pressure pores in a 200 meter well hole in Suzhou, China. The results showed unprecedented possibilities for monitoring soil decline, showing the possibility of access to breeds at any depth of the soil layers [25].

In recent years, Zhu et al. built a medium-size model of the soil slope, which was then subjected to a BOTDA sensor test as shown in Zhu et al. [26], where horizontal pressure distributions are obtained within the slope mass [26].

Klar et al. explored the possibility of investigating the incorporation/use of both BOTDA and Rayleigh scattering by relying on DOFS for induced tunnelling [27]. The authors' conclusions are that low-intensity sensors have been distorted according to land displacement and have also concluded that the DOFS technique provided an explanation for the results, which were only matched by a number of measurements of displacement without a millimeter [27].

6.3.4 HISTORICAL BUILDINGS

As is evident, there is a growing interest in observing historic buildings using fiber optic sensors, given their enormous importance in the cultural heritage of any society [28]. These sensors were used to study the special status of individual fiber optic sensor monitoring and have been widely practiced in this area. However, the same is not true for the DOFS case. We have referred to this type of application earlier regarding the application of Brillouin as a cheaper and effective supplement control. In this experiment, the efficiency of strain control was confirmed by the Brillouin DOFS system, even for reasonably low stress levels [28].

Another important example was described in Barrias et al. [29], where a fiber optic sensor was used to monitor a panel during the replacement of two pillars of the Sant Pau Hospital, the UNESCO World Heritage Site in Barcelona, as shown in Bastianini et al. [28]. The sensors were able to measure the effectiveness of the stress—and in this way, assess its structural stability and safety [29].

6.3.5 PAVEMENTS

To detect the flow of traffic, fiber optic sensors were used to design traffic flexibly [29]. This technique was developed using the micro-bending theory of optical fiber. When external force or compression is applied to fibers, the fibers bend over small semicircular filaments, thus breaking the center light of the fiber from the heart to the protective fiber layers, causing a decrease in light intensity [30].

Boby et al. [31] used fiber optic sensors embedded in road surface to detect vehicle weight. Also, Eckroth [32] suggested two methods in which a fiber optic temperature sensor (FOTS) can be embedded into the road surface (Figure 6.8) [32].

In one way, the FOTS was used horizontally in asphalt concrete, and in the other, it was used vertically. Cosentino et al. [33] have investigated the sensor work when placed in narrow vertical grooves. The study concluded that, compared with horizontally stabilized sensors, vertically stabilized sensors could avoid pressure concentrations from tires and had a longer life expectancy [33]. In addition, when the FOTS is placed in narrow vertical grooves, signal loss or light intensity is the result, where the tires load the surrounding pavement. On the other hand, accompanying groove movements are sufficient to cause loss of light that can be recorded using a computer on the side of the road as well as data acquisition systems [33].

For the purpose of not losing the signal due to weak light, Cosentino et al. [33] proposed placing vertically installed sensors near the pavement surface in a groove of about 3.2 mm width. The depth of the fine groove to improve the sensitivity was not precisely defined. Relying on lab and field data, as well as a series of models of limited elements and their connections, they suggested a depth of about 19 mm acceptable, and stated that depths more than this amount may cause premature structural damage to the pavement [33].

Bergmeister and Santa [34] installed the fiber optic sensor in the Koli-Iscarco Bridge neighborhood for the purpose of obtaining traffic loads. The sensors were installed as shown in Bergmeister and Santa [34], a double-refraction detector, embedded between two metal strips that were welded together. The results

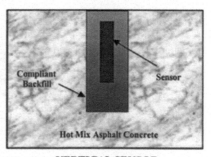

HORIZONTAL SENSOR   **VERTICAL SENSOR**

FIGURE 6.8   Fiber optic traffic sensors (FOTS) placement options [11].

demonstrated that the fiber optic sensor system in their study is reliable for several years [34].

Udd et al. have installed 28 traffic sensors using fiber optic sensors specifically designed at the Horsetail Falls, cutting holes in the Columbia River Gorge National Scenic Area in the United States [35]. The control system was tested by operating several cars with different weights of 10–18 km in the hour. In addition, they installed five other long-distance sensors on the I-84 highway to test the ability of these sensors to classify and respond to vehicles (Figure 6.9) [35].

The sensor system that has been developed can meet the requirements of traffic control because the amplitude of the signal is proportional to the weight and speed of the vehicle. It achieves a solution of less than 0.1 microwaves with a dynamic range of 400 microwaves at the 10 kHz sampling rate [35]. It also has the ability to distinguish between trailers, tractors, buses, and even traffic in adjacent passages in some cases [35]. It also showed the possibility of determining the driving direction by separating peaks and arranging their appearance in neighboring sensors. The results of the test clearly demonstrated the advantages of FBG sensors on conventional vehicle monitors [35].

6.3.6 TUNNELS

In general, for traffic safety purposes, a concrete lining is installed in the tunnel [1]. In order to follow a precise and effective approach, the structural health of the concrete lining must be monitored [2]. To achieve this, a single-mode optical fiber was connected by epoxy adhesive to a concrete tunnel surface with an internal diameter of 9.1 m. The strain along the optical fiber was measured by Brillouin optical correlation domain analysis (BOCDA) [1, 2]. Spatial accuracy of strain distribution is less than 100 mm. However, this precision may deteriorate depending on the measurement distance [1, 2]. From a practical point of view, consider the effect of temperature as well as optical polarization. In theory, using BOCDA, there is a trade-off between the measurable range and spatial location. Using this method, samples of the strain can be sampled at a rapid rate, and this approach is also very suitable for observing the dynamic strain in any arbitrary place [1, 2].

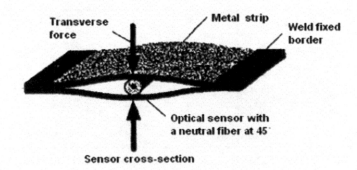

FIGURE 6.9 Fiber optic weight-in-motion sensors [11].

Of the important dangers that must be planned for, the non-occurrence of fire is one of the most important safety concerns in tunnels. Where it has occurred, it is very important to quickly detect it and locate it precisely [2]. In practice, the use of conventional sensors to detect fire is costly and inefficient. A distributed temperature sensor system based on temporal time measurement was used in Raman in China's Xuanwu Tunnel [1, 2]. The length of the tunnel is 1.6 km. A fire sensor system has been linked to the principle of fiber optic sensing using 30-year-old structures along the tunnel's crown, which is expected to be the hottest location, as fire causes hot air to rise. The stabilized system was tested by igniting a 50 cm diameter tank of diesel. The system succeeded in locating the fire in the very short time of 80 seconds [2].

6.3.7 EMBANKMENTS AND SLOPES

Internal pressure of dams and embankments is an important parameter for health evaluation [1]. With extremely high sensitivity, the interferometer is a candidate for measuring the displacement of a diaphragm under pressure. There were many pressure sensors developed based on extrinsic Fabry–Perot interferometer (EFPI) [2] as the displacement precision of EFPI can be less than 1 μm relative to the reference point. However, when the measurement device is switched off, the reference point is lost. So, the conventional configuration of EFPI may not be suitable candidate of long-term static measurement. A dual-EFPI pressure sensor that could specify a stable zero-point for every intermittent measurement was therefore developed [1, 2].

An important milestone for assessing the proper performance of both dams and bridges is the internal pressure parameter [1]. The important point to be made to measure health status is the high sensitivity of the instrument. On this basis, the interference scale is optimal for measuring the displacement of the diaphragm under pressure. Many conventional sensors are used to measure pressure on EFPI [2], and with acceptable accuracy. However, there are many reasons why traditional EFPI-based devices are not suitable for long-term measurement. Therefore, a double pressure sensor has been developed for EFPI [1, 2].

One of the most important points that must be monitored to maintain the health of important facilities such as dams and bridges is excessive displacement. For the purpose of effective monitoring of the originator, and to obtain sufficient results to measure this factor, the presence of many point sensors is required; this process is considered inefficient and uneconomical [2]. A fiber-optic-distributed sensor provides a good alternative for dams or bridges [2].

By using the stimulated Brillouin scattering measured by the BOTDA or the Brillouin optical frequency domain analysis (BOFDA) [2], the displacement of the strain distributed along the optical fiber is calculated. The BOFDA may be less expensive than the BOTDA. However, the maximum strain of common silica fiber is only about 5%, and this is not enough to monitor slope stability and soil creep [1, 3].

6.3.8 OTHER APPLICATIONS

Fuhr et al. [36] mentioned the possibility of applying the fiber optic sensors to the Winooski Dam for hydroelectric power generation in Vermont. The system has been developed using a multi-function fiber capable of sensing vibrations and pressures and integrating them to measure and monitor the water pressure of the dam, as well as monitoring vibration frequencies and their inputs to the power section of the dam. Using this type of sensor, a problem was found during the start of the hydro-power plant when the expected generator efficiency was not achieved [37]. Also, in Switzerland, the fiber optic sensors were applied to the long-term monitoring tunnel near Sargans. The sensors were manufactured using glass-fiber-reinforced polymer (GFRP) with integrated FBG sensors. This technique has been applied to measure distributed stress fields and temperature [38].

In contrast, in Italy a port dock was equipped with the fiber optic sensors system for continuous monitoring [39]. These sensors were used to measure the continuous movement of the pavement during dredging and docking. The sensors also measured the amount of change in bending at the horizontal and vertical levels. In addition, many data have been collected automatically and continuously since then [39].

In their application in monitoring the structural integrity of long pipelines, fiber optic sensors have demonstrated its high control and damage detection capacity for pipelines compared with conventional methods. It was used in Indonesia to monitor a 110 km pipeline to monitor and warn of damage caused by drilling equipment, theft, landslides, or earth movement [40]. By using the principle of the effect of met-ric interference, the vibration monitoring system detects the location of the fault by detecting changes in the scattered light properties resulting from the fiber pressure disorders. In fact, this system was able to observe a 50 km long optical fiber with a precision of 0.1, and was able to successfully detect damage to the pipeline at a spe-cific site, resulting from a landslide [41].

In Tennyson et al. [42], fiber optic sensors have been verified to monitor pipeline behavior and safety. Tests were carried out on pipe sections under various condi-tions in terms of internal pressure, axial pressure, bending, and local restriction. The results obtained demonstrate that the fiber optic sensors can monitor changes in loads and detect distortions and measure plastic strains after high pressure [42].

In Glisic et al. [43], the average breeds of two groups of piles were monitored under axial compression, pullout, and bending using fiber optic sensors. The study showed significant results on the behavior of the alkawam, soil properties, yawg coefficient of piles, probability of cracking, normal force distribution method, max-imum load probability in the case of axial compression and withdrawal tests, bend-ing distribution, horizontal displacement, and bend tests [43]. Lee et al. [44] also conducted a series of tests for the purpose of assessing the possibility of applying the fiber optic sensor system to heap devices. Through the tests, the researchers found that the cohort distribution of the models was evaluated from the strains mea-sured by FBG sensors comparable in size and direction with those obtained from conventional pressure gauges. Based on the results, the authors found the possibility of using this type of probe in the excavated columns and other types of cast-iron piles [44].

In addition to this, there is another important use of DOFS in Lan et al. [45], which can be used to monitor the loss of pre-stress in concrete beams. They pro-posed a new smart line combining the BOTDA and FBG sensors on one optical fiber integrated into a 5 mm fiber-reinforced polymer (FRP) iron. Various stress-tested

RC packets were tested with this thread, which was compared with more conventional sensors [45]. The results demonstrated the validity of this technique by monitoring the spatial distribution of pre-stress loss, as well as the chronological history of both construction and in-service phase [45].

REFERENCES

1. Deng, L., and Cai, C.S. Applications of fiber optic sensors in civil engineering. *Struct. Eng. Mech.* 2007, 25 (5), 577–596.
2. Barrias, A., Casas, J.R., and Herrero, S.V. Review of civil engineering applications with distributed optical fiber sensors. In: 8th European Workshop on Structural Health Monitoring (EWSHM 2016), Spain, Bilbao, 5–8 July 2016.
3. Annamdas, V.G.M. Review on developments in fiber optical sensors and applications. *Int. J. Mater. Eng.* 2011, 1 (1), 1–16.
4. Gupta, B.D. *Fiber Optic Sensors: Principles and Applications*, New India Publishing, New Delhi, 2006.
5. Gholamzadeh, B., and Nabovati, H. *Fiber Optic Sensors*. 2, 2008.
6. Wanser, K.H., and Voss, K.H., Crack detection using multimode fiber optical time domain reflectometry. In: Proceedings of the Distributed and Multiplexed Fiber Optic Sensors, SPIE, Bellingham, WA, 2294, pp. 43–52, 1994.
7. Habel, W.R. Fiber optic sensor in civil engineering: Experiences and Requirements. In: Proceedings of the SMART Structures: Optical Instrumentation and Sensing Systems, SPIE, Bellingham, WA, 2509, pp. 12–19, 1995.
8. Liu, H., and Yang, Z. Distributed optical fiber sensing of cracks in concrete. In: Proceedings of the Optical and Fiber Optic Sensor Systems, SPIE, Bellingham, WA, 3555, pp. 291–299, 1998.
9. Lee, D.C., Lee, J.J., and Kwon, I.B. Monitoring of fatigue crack growth in steel structures using intensity-based optical fiber sensors. *J. Intell. Mater. Syst. Struct.* 2000, 11, 100–107.
10. Wolff, R., and Miesseler, H.J. Monitoring of prestressed concrete structures with optic fiber sensors. In: Proceedings of the 1st Europa Conference on Smart Structures and Materials, Glasgow, pp. 23–29, 1992.
11. Deng, L., and Cai, C.S. Applications of fiber optic sensors in civil engineering. *Struct. Eng. Mech.* 2007, 25(5), 577–596.
12. Tennyson, R.C., Mufti, A.A., Rizkalla, S., Tadros, G., and Benmokrane, B. Structural health monitoring of innovative bridges in Canada with fiber optic sensors. *Smart Mater. Struct.* 2001, 10, 560–573.
13. Grossmann, B.G., and Huang, L. Fiber optic sensor array for multi-dimensional strain measurement. *Smart Mater. Struct.* 1998, 7, 159–165.
14. Bonfiglioli, B., and Pascale, G. Internal strain measurements in concrete elements by fiber optic sensors. *J. Mater. Civ. Eng.* 2003, 15 (2), 125–133.
15. Kenel, A., Nellen, P., Frank, A., and Marti, P. Reinforcing steel strains measured by Bragg grating sensors. *J. Mater. Civ. Eng.* 2005, 17, 423–431.
16. Casas, J.R., and Cruz, P.J.S. Fiber optic sensors for bridge monitoring. *J. Bridge Eng.* 2003, 8 (6), 362–373.
17. Fuhr, P.L., Ambrose, T.P., Huston, D.R., and Mcpadden, A.P. Fiber optic corrosion sensing for bridges and roadway surfaces. In: Proceedings of the SMART Structures and Materials: Smart Systems for Bridges, Structures, and Highways, SPIE, Bellingham, WA, 2446, pp. 2–8, 1995.
18. Fuhr, P.L., and Huston, D.R. Corrosion detection in reinforced concrete roadways and bridges via embedded fiber optic sensors. *Smart Mater. Struct.* 1998, 7, 217–228.

19. Casas, J.R., and Frangopol, D.M. Monitoring and reliability management of deteriorating concrete bridges. In: Proceedings of the 2nd Integração Workshop on Life-Cycle Cost Analysis and Design of Civil Infrastructure Systems, A. Miyamoto, Frangopol, D.M., eds., Yamaguchi Univ., Ube, Japan, pp. 127–141, 2001.

20. Iwaki, H., Yamakawa, H., and Mita, A., Health monitoring system using FBG-based sensors for a 12- story building with column dampers. In: Proceedings of the SMART Structures and Materials: SMART Systems for Bridges, Structures, and Highways, SPIE, Bellingham, WA, 4330, pp. 471–477, 2001.

21. Whelan, M.P., Albrecht, D., and Capsoni, A., Remote structural monitoring of the cathedral of Como using an optical fiber Bragg sensor system. In: Proceedings of the SMART Structures and Materials and Non-Destructive Evaluation for Health Monitoring and Diagnostics, SPIE, Bellingham, WA, 4330, pp. 471–477, 2002.

22. Bastianini, F., Corradi, M., Borri, A., and Tommaso, A. Retrofit and monitoring of an historical building using 'smart' CFRP with embedded fibre optic Brillouin sensors. *Constr. Build. Mater.* 2005, 19 (7), 525–535.

23. Shi, B., Xu, H., Chen, B., Zhang, D., Ding, Y., Cui, H., and Gao, J. A feasibility study on the application of fiber-optic distributed sensors for strain measurement in the Taiwan Strait Tunnel project. *Mar. Georesources Geotechnol.* 2003, 21 (3–4), 333–343.

24. Rajeev, P., Kodikara, J., Chiu, W.K., and Kuen, T. Distributed optical fibre sensors and their applications in pipeline monitoring. *Key Eng. Mater.* 2013, 558, 424–434.

25. Gue, C.Y., Wilcock, M., Alhaddad, M.M., Elshafie, M.Z.E.B., Soga, K., and Mair, R.J. The monitoring of an existing cast iron tunnel with distributed fibre optic sensing (DFOS). *J. Civil Struct. Health Monit.* 2015, 5 (5), 573–586.

26. Zhu, H.-H., Shi, B., Zhang, J., Yan, J.-F., and Zhang, C.-C. Distributed fiber optic monitoring and stability analysis of a model slope under surcharge loading. *J. Mt. Sci.* 2014, 11 (4), 979–989.

27. Klar, A., Dromy, I., and Linker, R. Monitoring tunneling induced ground displacements using distributed fiber-optic sensing. *Tunn. Undergr. Sp. Technol.* 2014, 40, 141–150.

28. Bastianini, F., Corradi, M., Borri, A., and di Tommaso, A. Retrofit and monitoring of an historical building using 'Smart' CFRP with embedded fibre optic Brillouin sensors. *Constr. Build. Mater.* 2005, 19, 525–535.

29. Barrias, A., Casas, J.R., Villalba, S., and Rodriguez, G. Health Monitoring of real 10 structures by distributed optical fiber. In: Proceedings of the Fifth International Symposium on Life-Cycle Civil Engineering, IALCCE'16, Held in Delft, Netherlands on October 16–19, 2016, 2016.

30. Grossman, B.G., Cosentino, P.J., Kalajian, E.H., Kumar, G., Doi, S., Verghese, J., and Lai, P., Fiber optic pore pressure sensor development. Transportation Research Record 1432, Transportation Research Board, Washington, DC, pp. 76–85, 1994.

31. Boby, J., Teral, S., Caussignac, J.M., and Siffert, M. Vehicle weighing in motion with fiber optic sensors. *Meas. Contr.* 1993, 26, 45–47.

32. Eckroth, W.V. Development and modeling of embedded fiber-optic traffic sensors. PhD Dissertation, Florida Institute of Technology, Melbourne, FL, 1999.

33. Cosentino, P.J., von Eckroth, W.V., and Grossman, B.G. Analysis of fiber optic traffic sensors in flexible pavements. *J. Transp. Eng.* 2003, 129, 549–557.

34. Bergmeister, K., and Santa, U. Global monitoring concepts for bridges. *Struct. Concr.* 2001, 2 (1), 29–39.

35. Udd, E., Kunzler, M., Laylor, M., Schulz, W., Kreger, S., Corones, J., McMahon, R., Soltesz, S., and Edgar, R., Fiber grating systems for traffic monitoring. In: Proceedings of the Health Monitoring and Management of Civil Infrastructure Systems, SPIE, Bellingham, WA, 4337, pp. 510–516, 2001.

36. Fuhr, P.L., and Huston, D.R. Multiplexed fiber optic pressure and vibration sensors for hydroelectric dam monitoring. *Smart Mater. Struct.* 1993, 2, 260–263.

37. Fuhr, P.L., Huston, D.R., Ambrose, T.P., and Barker, D.A. Embedded sensor results from the Winooski one hydroelectric dam. In: Proceedings of the Smart Structures and Materials, SPIE, Bellingham, WA, 2191, pp. 446–456, 1994.
38. Nellen, P.M., Frank, A., Bronnimann, R., and Sennhauser, U. Optical fiber Bragg gratings for tunnel surveillance. *Proc. SPIE* 2000, 3986, 263–270.
39. Inaudi, D., Del Grosso, A., and Lanata, F. Analysis of long-term deformation data from the San Giorgio Harbor pier in Genoa. In: Proceedings of the Health Monitoring and Management of Civil Infrastructure Systems SPIE, Bellingham, WA, 4337, pp. 459–465, 2001.
40. Fernandez, M.L., Tapanes, E.E., and Zelitskaya, P.V. Pipeline hydrocarbon transportation: Some operating concerns and RD trends. In: Proceedings of the 1st Integração Pipeline Conference, ASME OMAE, 1, pp. 95–102, 1996.
41. Li, H., Li, D., and Song, G. Recent applications of fiber optic sensors to health monitoring in civil engineering. *Eng. Struct.* 2004, 26, 1647–1657.
42. Tennyson, R.C., Morison, W.D., and Manuelipillai, G. Intelligent pipelines using fiber optic sensors. In: Proceedings of the Smart Structures and Materials: Smart Sensor Technology and Measurement System, SPIE, Bellingham, WA, 5050, pp. 295–304, 2003.
43. Glisic, B., Inaudi, D., and Nan, C. Pile monitoring with fiber optic sensors during axial compression, pullout, and flexure tests. *Transp. Res. Rec.* 2002, 1808, 11–20.
44. Lee, W., Lee, W.J., Lee, S.B., and Salgado, R. Measurement of pile load transfer using the fiber Bragg grating sensor system. *Can. Geotech. J.* 2004, 41 (6), 1222–1232.
45. Lan, C., Zhou, Z., and Ou, J. Monitoring of structural prestress loss in RC beams by inner distributed Brillouin and fiber Bragg grating sensors on a single optical fiber. *Struct. Control Health Monit.* 2014, 21, 317–330.

7 Fiber Optic Sensors for Biomedical Applications

7.1 INTRODUCTION

Fiber optic technology provides a safe and efficient way to receive and collect light into and from the tissue area of interest; it has been clinically employed since the 1960s [1–5]. This chapter discusses and reviews recent advances in fiber optic sensor technology in the field of biomedicine.

7.2 BIOMEDICAL SENSORS

Biomedical sensors are among the most efficient and reliable in terms of results compared to traditional instruments for measuring biological, chemical, and physiological factors in the living body [6] are classified as life-saving devices, with their precise potential for obtaining physiological information and their high ability to address them [7]. The high sensitivity that should be provided when dealing with the living body in terms of measurements, especially in terms of the necessities of the need for new devices that are small, intelligent, reliable, sensitive, and biologically compatible and remote control in nature is provided by biomedical sensors [8, 9].

These devices have the unique ability to select the required parameter without interference from other parameters. These devices include direct/indirect, contact/remote, gaseous/non-acute, real-time/hard, sensory/operator [10]. In general, an olfactory sensor usually consists of two main components: the sensing component that measures and the energy converter that converts the measurement to an output signal [11, 12]. This is shown in Figure 7.1 [13].

The rapid acceleration in recent scientific development has led to the emergence of wearable sensor technology, which has been used in the medical field for continuous observation, which has revolutionized the progress of this area greatly. The high need for reliable and high-precision devices that continuously monitor living-body devices has led to the development of the microelectronics industry with improved signal processing techniques, resulting in the production of effective bioreactors. Hence a great demand on these vital devices as a result of the need for good health care [13].

Currently, biomedical sensors are widely used to measure and monitor many important parameters such as body temperature, cholesterol, heart rate, blood flow velocity and pressure, respiratory level, oxygen ratio, blood micronutrient level, and daily exercise activities (Figure 7.2) [7–10].

In addition, biomedical sensors are used in complex surgical procedures, treatment purposes, and accurate diagnosis. It also helps reduce the incidence and increase the prevention of serious diseases by diagnosing them using low-intervention techniques

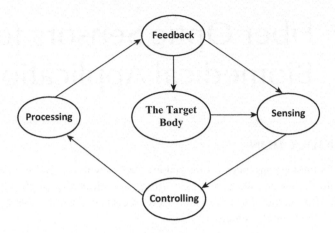

FIGURE 7.1   Biomedical instrumentation system.

such as magnetic resonance imaging (MRI), X-ray imaging, computed tomography (CT) scan, ultrasound imaging, endoscopy, and spectrum [14].

7.3   BIOMEDICAL APPLICATIONS

Fiber optic sensors are used in many applications as a result of their characteristic properties, resulting in effective use in many branches of science and engineering, ranging from measurement and control of temperature, pressure, strain, fluid level, pH, chemical analysis, etc. [2, 3]. They are also used to measure the refractive index of the material for the purpose of studying the molecular structure and identification of organic compounds (Figure 7.3) [4, 5].

7.3.1   GLUCOSE SENSOR

Previously, UV rays were used in addition to other techniques for sensing purposes [15, 16]. Currently, they have been replaced with a fiber optic sensor method based on the pH meter [17]. In this method only one broadband band is measured, allowing the system to adapt to the infrared laser diode system, thus providing the possibility of minimization and increased portability. Brown et al. have adjusted the sensor by using fixed glucose oxidase on the surface of the polyaniline polymer to predict glucose concentrations [18].

7.3.2   LAMINATE CURE ANALYSIS

The ability to monitor interactions in a hostile environment becomes easier by using sensors with small dimensions and sufficient durability. Because of its small size, fiber optic sensors can be introduced into the autoclave through the thermal calibration port, and thus can continuously monitor the progress of the reactions as a function of the operating conditions [19]. Druy et al. take advantage of this feature to monitor ongoing processes in many industries and applications [20].

7.3.3 Protein Analysis

Fourier-transform infrared spectroscopy (FTIR) with fiber optic investigation is one of the most useful ways to analyze proteins. High-quality spectra can be obtained from low concentrations of the analyzing in a variety of environments without interference. Usually, globular proteins exhibit regions of secondary structure, including alpha helices, P-sheets, turns, and non-ordered regions [21]. These harmonic entities contribute to the infrared spectrum. In addition, FTIR combined with optical fiber probes is useful, especially in the study of soluble proteins whose structure cannot be determined using X-ray diffraction or infrared spectroscopy [22].

7.3.4 Dosage Form Analysis

Drissie et al. reported on the possibility of applying the fiber optic probe to monitor the quality of the pharmaceutical industry [23]. In this system, they were able to use the fiber optic sensor to quantify the content of a number of solid pharmaceutical forms containing ibuprofen, and powders containing analgesic benzene analog of cetrimide. A team from Purus Wellcome took another step forward, conducting identification tests on tablets through the plastic wall for wart packaging [24] to distinguish between coated and unpainted discs and between active and placebo forms. The results were good, and the technology met all requirements to confirm the identity test prior to use in a clinical trial [25].

7.3.5 Drug Identification

Other applications of fiber optic sensors were performed by Arens et al.; they proposed an analysis of the systematic determination of toxicity using high-performance thin layer chromatography to detect drugs in biological samples. Using this method, parallel chromatography was allowed by identifying drugs and comparing their ultraviolet spectra with data obtained from the library as reference spectra [26].

7.3.6 Determination of DNA Oligomers

Kleinjung et al. have demonstrated method to immobilized DNA targets using a fiber optic sensor [24]. In this method, 13-mer oligonucleotides were attached to the core of a multi-mode fiber and the complementary sequence was identified by using a fluorescent double-stranded specific DNA ligand [27]. For the purpose of distinguishing between bound and unbound species, the evanescent field was used. The process begins by freezing the DNA oligomer template by direct coupling to the activated sensor surface or using the avidin biotin bridge to detect the mismatch of the single base in the target sequence [27].

7.3.7 Pesticide Detection

Rajan et al. suggested a method for detecting the organic phosphate pesticide by an industry characterized by fiber optic surface plasmon resonance (SPR) based on the fiber optic sensor [25]. In which the acetylcholine enzyme (ACHE) is prepared to prepare the probe on the coated silver core of the plastic silica fiber (PCS), the detection is based on the principle of the competitive bonding of the substrate to the ACHE enzyme [28]. For the fixed concentration of the substrate, with the increased concentration of the pesticide, the SPR wavelength will decrease; this increase in the amount of pesticide caused a comprehensive reduction in sensitivity [28].

7.3.8 Effluent Monitoring

Krska et al. have demonstrated the environmental risks of pharmaceutical manufacturers using chlorinated hydrocarbons [29]. These risks are due to the inclusion of chloro-hydro-bronate on their strongest absorption ranges by using fiber optic sensors so that the crystal silver halide fibers have a light value that can be sensed. In order to perform quantitative measurements, 10 cm fiber clusters were combined with FTIR, and samples were monitored. In contrast, another study revealed a comparative analysis of ethane tetrachloride samples and water wastes that showed a good correlation with standard gas chromatography techniques [29].

7.3.9 Other Applications

The use of fiber optic sensors is not limited to water determination by near-infrared spectroscopy [30] but is also used to identify penicillamine in pharmaceuticals and human plasma through capillary electrophoresis with deep fiber optics from diode light [31]. In addition, fibrous lasers are used in many other applications, such as military, biological, biomedical, and air-sensitive gases [32, 33].

Many of the applications can be performed using the Raman sensor for filtered fiber, such as measuring high levels of organic solvents in soil, aquifers, monitoring processes and chemical activities of petrochemicals and distillation products, monitoring polymer interaction reactions at the site, and many other factors [36].

Other applications are performed in spectroscopy. In order to analyze the composition of a substance that cannot be placed in the spectrometer itself, it can be measured by optical beams by transferring light from a spectrometer to a substance. In this case, the spectrometer analyzes the material by reversing light through it. Also using fiber, a spectrometer can be used to study objects that are too large to be placed inside, gases, or interactions in pressure vessels [37].

7.4 SUMMARY

This chapter reviewed the use of optical fiber Bragg gratings for biomedical sensor applications (i.e. glucose sensor, laminate cure analysis, protein analysis, dosage form analysis, drug identification, determination of DNA oligomers, pesticide detection, effluent monitoring, and others). Also, the technology for measuring the biological, the chemical, and the physiological parameters of the body has been reviewed.

REFERENCES

1. Kapany, N.S. *Fiber Optics. Principles and Applications*, Academic Press, New York, NY, 1967.
2. Binu, S. Calibration of accelerometers by using an extrinsic fiber optic probe. *Microw. Opt. Technol. Lett.* 2007, 49, 2700–2703.
3. Culshaw, B. Optical systems and sensors for measurement and control. *J. Phys. E: Sci. Instrum.* 1983, 16, 978–986.
4. Karrer, E., and Orr, R.S. A photoelectric refractometer. *J. Opt. Soc. Am.* 1946, 36, 42–46.
5. Ross, I.N., and Mbanu, A. Optical monitoring of glucose concentration. *Opt. Laser Technol.* 1985, 17, 31–35.
6. Mendez, A. Biomedical fiber optic sensor applications. In: Conference 2015 Optical Fiber Communications Conference and Exhibition (OFC), Los Angeles, CA, 2015, pp. 1–64.
7. Frantisek, U., and Jozef, C. Current trends in photonic sensors. In: Proceedings of the 2014 24th International Conference 2015 Radioelectron, Bratislava, Slovakia, 2014, pp. 1–6.
8. O'Donnell, M. Biomedical instrumentation and design, BME/EECS 458. pp. 1–10. Available from: http://www.peb.ufrj.br/cursos/cob783/chapter1_notes.pdf.
9. Cote, G.L., Lec, R.M., and Pishko, M.V. Emerging biomedical sensing technologies and their applications. *IEEE Sensors J.* 2003, 3 (3), 251–266.
10. Roriz, P., Frazão, O., António, B., Lobo-Ribeiro, B., Santos, J. L., Simõesa, José A. Review of fiber-optic pressure sensors for biomedical and biomechanical applications. *J. Biomed. Opt.* 2013, 18(5), 050903.
11. Jones, T.P., and Porter, M.D. Optical pH sensor based on the chemical modification of a porous polymer film. *Anal. Chem.* 1988, 60, 404–406.
12. Zhang, S., Tanaka, S., Wickramsighe, Y.A., and Rolfe, P. Fiber-optical sensor based on fluorescence indicator for monitoring physiological pH values. *Med. Biol. Eng. Comput.* 1995, 33, 152–156.
13. Takpor, T.O., and Agboje, O.E. Advances in optical biomedical sensing technology. In: Proceedings of the World Congress on Engineering 2016 Vol I WCE 2016, London, UK, June 29–July 1, 2016.
14. Ge, Z.F. Fiber optic near and mid infrared spectroscopy and clinical applications. *Diss. Abstr. Int.* 1995, 55, 3855.
15. Brown, C. Near and mid infrared chemical and biological sensors. *Proc. SPIE Int. Sot. Opt. Eng.* 1995, 2506, 243–250.
16. Druy, G., and Stevenson, W.A. Embedded optical fiber sensors for monitoring cure cycles of composites. In: Proceedings of the ADPA/AIAA/ ASME/ SPIE, IOP Publishers, UK, 1992.
17. Druy, G., and Stevenson, W.A. Mid-IR tapered chalcogenide fiber optic attenuated total reflectance (ATR) sensors for monitoring epoxy resin chemistry. *SPIE Int. Soc. Opt Eng.* 1993, 2089, 114–120.
18. Haris, P.I., and Chapman, D. Does Fourier transform infrared spectroscopy provide useful information on protein structures? *Trends Biochem. Sci.* 1992, 17, 328–333.
19. Mantele, W. Reaction induced infra-red spectroscopy for the study of protein function and reaction mechanisms. *TIBS* 1993, 18, 197–202.
20. Dreassi, E., Ceramelli, G., Corti, P., Massacesi, M., and Perruccio, P.L. Quantitative Fourier transform near infrared spectroscopy in the quality control of sold pharmaceutical formulations. *Analyst* 1995, 120, 2361–2365.

21. MacDonald, B.F., Gemperline, P.J., and Boyer, N.R. A near infrared reflectance analysis method for the non invasive identification of film coated and non film coated, blister packed tablets. *Chim. Acta* 1995, 310, 43–51.

22. Blanco, M., Coello, J., Iturriaga, H., Maspoch, S., and Russo, E. Control analysis of a pharmaceutical preparation by near infrared Dempster reflectance spectroscopy. A comparative study of a spinning module and a fiber optic probe. *Anal. chim. acta* 1994, 298, 183–191.

23. Ahrens, B., Blankenhorn, D., and Spangenberg, B. Advanced fibre optical scanning in thin layer chromatography for drug identification. *J. Chromatogr. B* 2002, 772, 11–18.

24. Kleinjung, F. Bier, F.F., Warsinke, A., and Scheller, F.W. Fibre-optic genosensor for specific determination of femtomolar DNA oligomers. *Anal. Chim. Acta* 1997, 350, 51–58.

25. Rajan, C.S., and Gupta, B.D. Surface plasmon resonance based fiber optic sensor for the detection of pesticide. *Sens. Actuators B* 2007, 123, 661–666.

26. Krska, R., Rosenberg, E., Taga, K., Kellner, R., Messica, A., and Katzir, A. Polymer coated silver halide infrared fiber as sensing devices for chlorinated hydrocarbons in water. *Appl. Phys. Lett.* 1992, 61, 1778–1780.

27. Krska, R., Taga, K., and Kellner, R. Simultaneous in situ trace analysis of several chlorinated hydrocarbons in water with an IR fiber optical system. *J. Mol. Struct.* 1993, 294, 1–4.

28. Krska, R., Taga, K., and Kellner, R. New IR fiber optic chemical sensor for in situ measurements of chlorinated hydrocarbons in water. *App. Spect.* 1993, 47, 1484–1487.

29. Blanco, M., Coello, J., Iturriaga, H., Maspoch, S., and Rovira, E. Determination of water in ferrous lactate by near infrared reflectance spectroscopy with a fiber optics probe. *J. Pharm. Biomed. Anal.* 1997, 16, 255–262.

30. Yang, X., Yuan, H., Wang, C., Su, X., Hu, L., and Xiao, D. Determination of penicillamine in pharmaceuticals and human plasma by capillary electrophoresis within column fiber optics light emitting diode induced fluorescence detection. *J. Pharm. Biomed. Anal.* 2007, 45, 362–366.

31. Richter, D., Fried, A., Wert, B.P., Walega, J.G., and Tittel, F.K. Development of a tunable mid ir difference frequency laser source for highly sensitive airborne trace gas detection. *Appl. Phys. B* 2002, 75, 281–288.

32. Greenwald, J., Rosen, S., Anderson, R.R., Harrist, T., MacFarland, F., Noe, J., and Parrish, J.A. Comparative histological studies of the tunable dye (at 577 nm) dye laser and argon laser: The specific vascular effects of the dye laser. *J. Invest. Dermatol.* 1981, 77, 305–310.

33. Lyon, R.E., Chike, K.E., and Angel, S.M. In situ cure monitoring of epoxy resins using fiber-optic probe. *J. Appl. Polym. Sci.* 1994, 53, 1805–1812.

34. Angel, S.M., Vess, T., Langry, K., and Kyle, K.T. In Proceedings of Symposium on Chemical Sensors II. *J. Electrochem. Soc.* 1993, 93, 625.

35. Garrison, A.A., Moore, C.F., Roberts, M.J., and Hall, P.D. Distillation process control using Fourier tranform Raman spectroscopy. *Process Control Qual.* 1992, 3, 57–63.

36. Mosheky, Z., Melling, P.J., and Thomson, M.A. In situ real time monitoring of a fermentation reaction using a fiber optic FT-IR probe. *Spectroscopy* 2001, 16, 15. Available from: http://www.remspec.com/pdfs/SP5619.pdf.

8 Fiber Optic Sensors for Military Applications

8.1 INTRODUCTION

Over the last few years, there has been increasing commercial interest in the optical communications sector, coupled with a revolution in the field of research that has resulted in significant progress in the development and application of fiber optic sensors [1]. During this massive boom, the military field had little space compared to other civilian applications. Fiber optic technology can provide significant benefits to the military, especially in the fields of communications and sensing [1]. The characteristics enjoyed by fiber optics make it an important option for military communications.

The advantages of fiber optics make it an important option for military communications. They are resistant to corrosion and electromagnetic interference (EMI) and are therefore very suitable for conditions in which military zones are characterized as chemical or nuclear environments. It also produces no electric fields, making it invisible to the enemy's eyes. In addition to communications, the use of this technology to perform other military tasks such as guidance, sensing, and detection can be greatly beneficial [1].

8.2 FIBER OPTICS FOR MILITARY APPLICATIONS

As a result of its unique advantages, the use of fiber optic technology in military communications systems will provide many benefits. First, it is immune to EMI and radio-frequency interference (RFI) noise; second, it can support very long distances before the repeater is necessary; third, it offers a very large bandwidth. It can be used in audio, video, and data applications, and its cost is low because many domestic and foreign manufacturers can now produce high-quality fiber cables, connectors, and transceivers [2].

Currently, commercial markets can provide products that meet the basic needs of military network design. Where it can provide special products such as specialized military links such as straight tip (ST) with high tension springs to ensure that no shock or vibration occurs. It can also provide other variants of the ST fiber link, which features a locking mechanism that prevents optical interconnection in high vibration or vibration environments [2].

Of the military-type products, the fiber optic connectors of the IP designs have differences that make them even stronger when exposed to shocks and vibrations. The usual stress rate on these connections is more than 250 Newton (56 lbs), which is more than 50% better than standard designs in conventional applications, a commercial fiber conductivity mitigation technology. This additional protection ensures that the cable pool will remain in an exemplary connection during field use [2].

Many companies provide fiber optics for military/industrial applications, which are characterized by an armored steel exterior, in addition to special designs for the design of flex cables, simple design, and double design [1].

8.3   SOME FIBER OPTICS SENSORS' MILITARY APPLICATIONS

Much research has been done on the possibility of using fiber optics to improve many military operations. Many of these studies examine how optical fibers are used in weapons systems, observation, optical computing, onboard vehicles, the transmission of information, and in various sensor techniques [1, 2].

The use of optical fibers in military fields is equivalent to their use in commercial applications. These uses include the characteristics of optical fiber communication systems that are of paramount importance in the military field rather than in commercial applications such as electromagnetic immunity, relative security of eavesdropping, long distance spans without electronic repeats, and the relatively low cable weight [2].

Additional requirements that must be met in optical fibers to suit military uses include the possibility of working for a wide range of temperature during operation and storage; high tolerance and vibration, strain and high shock, and other mechanical stresses; and system performance strength [2].

8.3.1   COMMUNICATIONS

The most important characteristic of military applications compared to other applications is the subject of immunity and protection of the nature of information and techniques used. Therefore, from a purely security point of view, optical fiber is very suitable because it does not produce any electromagnetic fields; it is therefore impossible to listen to them. For these reasons, fiber optics are an indispensable option for all types of communications, both long and short. It is also one of the most secure systems for unencrypted communications [2].

With regard to Air Force Photovoltaic Communications, a special system has been established by the Hughes Microelectronics Systems Department to detect any breach or intrusion. The system connects two computer systems and local networks, and has the ability to transmit data up to 12 Mbps up to 1.5 kilometers. As a mean of protecting information, this system will automatically cease if the signal it is transmitting is intercepted [2]. The work is being conducted by researchers at the Aviation Development Center in Rome at Hanscom Air Base to increase the system's capacity to 100 Mbps over 35 km. A tactical version of this system can be used if fiber optic connectors can be easily set up and downloaded [2].

Fiber optic networks have other systems in place to install their units as required by circumstances, so these units in the field can simply "connect" to the network to establish a secure communication link. Thus, the military can use fiber optics as needed, and can provide temporary short-range communications, depending on the geographical area required. In addition, optical fibers are simply stored on the ground by a few trained persons or transported for a certain distance by using special vehicles to transport fiber optic cable [2]. A set close to the landing point can connect the fiber with a small remover switch, and a secure communication link will be set up [2].

8.3.2 WEAPON SYSTEMS

Military equipment companies are racing among themselves to develop communication systems and equipment to provide secure communication paths. Currently, the only real-time weapon system developed is the fiberoptic guided missile (FOG-M). The system has an average range of 20–40 km, and has a lock system after its launch, which is used against helicopters and ground vehicles. [3] The Navy fiber optic gyroscopes (FOGs), a fiber-guided weapon, is also proposed. In addition, Hughes proposes a fiber-guided weapon for anti-medium advanced weapons (AAWS-M). In addition, there is a handheld weapon, a short-range weapon (2 km), which is used primarily against tanks [4–6]. Both FOG-M/S and Hughes AAWS-M concepts are deadly weapons, and a man is used in the ring to get guidance. In all of these weapons, the fiber optic cable that is propelled from the rocket during the flight is used to transfer a picture to the gun from the camera (TV or infinite impulse response [IIR]) in the nose of the missiles [2, 5, 6].

Typical optical fibers used in military systems and applications are monocrystalline fibers with a silica nucleus dissected with germanium or phosphorus and pure silica cladding with a clear ultraviolet polymer coating used to protect fibers from damage [2]. The outer diameter of the fibers is 200–250 microns. These weapons have become available and used by the optical fiber data link. One of the main disadvantages of a copper cable is its lack of bandwidth necessary to transfer a television image from the camera in the rocket to the CRT in the cannon station as in fiber optics. For this reason, the current man in TOW & Dragon systems requires the gun to hold the cross beam on the target until the missile strikes the target [2].

In addition, the systems that use the copper cable have a short range due to their weight and large area so that the missile can carry the appropriate length. On the other hand, in order to operate a fiber optic weapon, the fiber must speed up without breaking. Several tests were conducted to determine the speed of fiber loading and were found to be up to 600 ft/s [7, 8], which is the maximum theoretical speed of thrust [9]. The FOG-M program is also developed in the Forward Area Air Defense System (FAADS) program where the non-line-of-sight (NLOS) component is primarily used against helicopters [2, 8, 9].

This type of missile has the advantage of carrying a warhead in tandem and is widely used to target tanks. [10] In a war environment, it is estimated that about 25% are used against helicopters and 75% are used against ground targets. It is also possible that two high-powered multi-purpose rocket-propelled grenade launchers (HMWWVs) 8 and a multi-rocket MLRS launcher with 12 or 24 missiles will be used.

The first of these missiles was 16,000 rockets; more than 8,000 were directed by television and the rest were directed by the IIR. Because of the cost of electrons, the cost of the rocket is expected to be almost low compared to other technologies [11].

As carried out at the Marine Weapons Research, Development, Testing and Assessment Facility in Lake China, California, A-7 Corsairs launched fiber optic weapons while flying [12].

One of the characteristics of this rocket is that it is very close, if not identical, to the current FOG-M concept. The main difference is the requirement of a second pay-off on the aircraft to achieve sufficient capability to launch the missile and then turn and fly in any direction required according to the direction of the target and even in the reverse movement of the rocket itself [13]. In addition to being a weapon, it can be used as a means of monitoring unmanned aerial vehicles. Fiber optic technology is one of the most important tactical applications used by aircraft to drop buoys in the ocean using fiber optic missiles [13].

These networks are connected by fiber optic or radio links to the aircraft, to the ship, or to the ground base, so that the missiles can be fired whenever the target moves, whether a ship or an aircraft, within the range allowed [13]. Queuing for this type of system can be radar, satellite, or aircraft. This technique can be used if both aircraft and ships are lost to a large part of the ocean. In addition, another weapon system is a set of buses and submunitions using fiber optic connections [13]. Also, the systems used in the military formations are Precision Deep Attack Missile System (PDAMS), a ballistic missile bus very similar to the Army Tactical Missile System (ATACMS), which carries submunitions connected to the bus with optical fiber cables [13].

In these units, the bus is connected to the operator by a wireless frequency link. This system (PDAMS) is used to target and destroy tactical ballistic missiles (TBMs) within less than 5 minutes from the outbreak of hostilities. So they have the ability to target rocket launchers and destroy them before they can fire any rockets [14]. In addition, the ability to use a delivery vehicle with the ability to identify targets can produce tactical superiority and significantly improve the current concept of war deterrence. This type of bus is fiber optic and carries smart submunitions to the target location [15].

8.3.3 SURVEILLANCE/SENSORS

Fiber optic sensors have a large and very diverse range of sensing applications because of their extremely unique properties and their high ability to observe a wide range of physical parameters.

As an example of commercial sensing applications, photovoltaic sensors are used in the petroleum industry within the oil wells for their ability to withstand extremely hot environments [14]. In addition, the Defense Nuclear Agency (DNA) supports the development of pressure sensors for nuclear measuring devices. Fiber optic pressure sensors seem to have ranges greater than 10 kbarts and have been shown to be able to track pressure pulses at very high precision [15].

In addition to the potential use of optical fiber as part of the military technology, it can also be used to monitor war zones effectively, as it can transmit information, whether data, pictures, or video quickly and accurately [15]. It also provides itself well for the battlefield environment, being very small and sensitive to many parameters, such as heat, pressure, strain, and others. Recent years have seen the use of optical fibers as sensors in battlefields, where a pressure sensor buried in a road or a seismic sensor set along the way can determine the number of vehicles passing through a point [16].

Another possible use of optical fibers is to monitor unmanned aerial vehicles through a link between the vehicle and the ground station. The data link can also be used to transmit information and guidance to the vehicle and images, whether visual, infrared, or radar, from the vehicle to the ground station [17]. Also, fiber optic sensors can be placed on light aircraft, and the output via fiber is transferred to the ground station. It can also be used on the vehicle as a lizzy router as a weapon such as Copperhead. The basic advantages of using a fiber optic link are immunity to jamming, EMI, and secret operation [17].

8.3.4 ABOARD VEHICLES

The many advantages of optical fiber make it suitable for many applications, especially for military and commercial vehicle manufacturing applications. Land, sea, and air vehicles can benefit from the advantage of light weight to reduce the total mass of these compounds [18]. The focus is on the design of lightweight aircraft to provide fuel and absorb radar rays to disappear from the eyes of the enemy. Not only is optical fiber much lighter than copper cables, it has immunity against lightning strikes, making electrical shielding unnecessary [18].

Optic fiber is used as a conduit for the transmission of optical signals from the flight control computer in the gondola to the rear control surfaces. The fiber optic path lengths are fully proportional to this application due to the balloon's path [19]. According to the Director of Planning of the National Aeronautics and Space Administration, using a large amount of optical fiber on board spacecraft is very important because of the urgent need to reduce the total weight of vehicles [20].

The use of long-distance optical fiber to connect elements of multi-stage radar platforms is indispensable because of the large savings that will be saved in weight [21]. Radar pads contain many elements connected to each other by optical links, which means significantly reducing the weight of the aircraft compared to the use of copper cables. In addition, to make full use of fiber optics in phase-stage radars, high-frequency waves must be transmitted in fibers [21]. Fiber optic cables can also be used to connect the elements of the progressive array if laser beams and high modulation rates can be developed [22–25].

What is required of these devices is to work within microwaves and millimeter waves used in radar systems. The use of optical fiber cables and different lengths to connect between the different elements of the intermediate matrix and the phase switch will significantly eliminate the need for an expensive electronic unit in each element to provide phase change [26]. Because only one phase shifter must be used, with the replacement of the longer fiber cables, the optical fiber conversion system will be much lighter. This is of great importance to airborne radar platforms. A longitudinal array antenna with four elements connected to fiber optic cables within the 2–3 GHz range has already been developed to reflect this concept [26]. Fiber optic gyroscopes have revolutionized navigation, just as gyroscopes did years ago. They are characterized by many advantages such as small size, light weight, low energy requirements, hardness, and modest cost. They also have the ability to compete for low and medium-precision applications [27].

The Strategic Defense Initiative has been studying FOGs for use on a space interceptor. Litton has also developed a three-axis failure measurement unit [28]. The same physical foundations were also used for the FOG design as it is used to design a circular laser gyroscope [28].

Because optical fibers have the ability to achieve a very long path length in a small area using a multi-grading file, the FOG type has been highly interested. Varo Incorporated has developed a laser-based laser warning system for onboard vehicles. This system operates in a wavelength range of 400 to 1000 nm and gives 90 degrees of coverage [29].

This system alerts the operator when his car is illuminated by laser. This sensor has been used in air, sea, and land vehicles for its efficiency and effective ability to cover fully. Maximum EMI levels, electrical problems, floods, and the corrosive environment make fiber a natural option for transporting information on ships [29, 30].

Since 1982, the U.S. Navy has introduced fiber optic technology to work within its configurations through the Aegis cruiser, wherein the fiber optic damage control system has been developed. Fiber optic sensors were also used to detect many parameters, such as smoke detectors, temperature, temperature change rate, and fire and flood control. [30] Research is also ongoing to find a new and highly exciting concept under development—the incorporation of fiber optic sensors into motor structures for the performance of health surveillance [30, 31].

This type of sensor can also be used within the vehicle's internal structures to monitor the overall structure of the chassis and its suitability to withstand the flight, to determine faults, and to inform the crew of a problem; or a central computer may initiate an automatic corrective action. This network of sensors can also be used to alert maintenance teams of problems or the need for repairs. This type of sensor network inside the airframe can be very convenient for those concerned about the structural integrity of the aging transport fleet, whose fears have been exacerbated by the likelihood of failure on board commercial aircraft in recent years [32].

This type of sensor represents the nervous system of many applications because it acts as the carrier nerve of the signal; as it senses the various parameters in different areas and because the focus is largely on the use of vehicles as part of the structure of pregnancy, one of the problems experienced by these vehicles is a failure with no prior evidence. Thus, the monitoring system may alleviate this concern by providing a warning of possible failure [33].

8.4 SUMMARY

In this chapter, we reviewed the use of the optical fiber Bragg grating for military applications (i.e. weapon systems, surveillance systems, optical computing systems, and aboard vehicles). Also, the use of fiber optics for military communications to transmit optical signals from the flight control computer in the gondola to the control surfaces at the ship's rear, the use of optical fiber path lengths to link the radar elements of an airborne phased-array radar platform, and others.

REFERENCES

1. Benzoni, J.F., and Orletsky, D.T. *Military Applications of Fiber Optics Technology*, The RAND Corporation, Santa Monica, CA, 1999.
2. McNeil, P. *Specialized Cabling for Military Applications*, L-com Global Connectivity, North Andover, MA.
3. Culshaw, B. Fiber optics in sensing and measurement . *IEEE J. Sel. Top. Quantum Electron.* 2000, 6 (6), 1014–1021.
4. Elster, J., Trego, A., Catterall, C., Averett, J., Evans, M., Jones, M., and Fielder, B. Flight demonstration of fiber optic sensors. In: SPIE - Smart Sensor Technology and Measurement Systems, March 2003.
5. Kaminow, I. P., Marcuse, D., and Presby, H. M. Multimode fiber bandwidth: Theory and practice. *Proc. IEEE* 1980, 68 (10), 1209–1213.
6. Klass, P.J. Firms Research Fiber-Optic Gyros as Successors to Ring-Laser Systems. *Aviation Week & Space Technology* 13 February 1989, 79–85.
7. Klocek, Paul, ed. Infrared optical materials and Fibers V. In: Proceedings of the International Society for Optical Engineering, Vol. 843, No. 20 August, 1987.
8. Landauer, R. Dissipation and noise immunity in computation and communication. *Nature* 1988, 335 (6193), 779–784.
9. Lemrow, C.M., and Reitz, P.R. Single mode optical waveguide specifications. *Telecommun. Mag.* 1984, 18 (5), 76–78.
10. Lewis, N., and Moore, E.L., (eds), Proceedings of the International Society for Optical Engineering, 840, 1987 Fiber optic systems for mobile platforms.
11. Li, T. Advances in Lightwave systems research. *AT&T Tech. J.* 1987, 66, 5–18.
12. Tingye Li, T. Structures, parameters, and transmission properties of optical fibers. *Proc. IEEE* 1980, 68, 1175–1180.
13. Lines, M.E. The search for very low loss fiber-optic materials. *Science* 1984, 226, 663–668.
14. Little, W.R., Otto, D.C., and Denier, C.A., *A New Approach to Sensors for Shipboard Use, SPIE*, 840, 1987.
15. Mason, V.A., (ed.) Mayo, john S., materials for information and communication,. *Telecom. Rep.*, Vol. 55, No. 8. *Scientific American*, 1986, 253, 59–65, p. 20.
16. Nicholson, P.J., (ed.) Telecommunications news. *Telecommunications* 1989, 23, 11.
17. Nordwall, B.D. Industry, defense pursue development of learning, adaptive neurocomputers. *Aviation Week and Space Technology*, 14 November 1988, 101–103.
18. Ohnsorge, H. Introduction and overview of broad-band communication systems. *IEEE J. Sel. Areas Commun.*, Vol. SAC-4 1986, 4, 425–428.
19. Optical integrated circuits, *International Defense Review*, G. Sundaram, (ed.), 21, 320, 1988.
20. Pandhi, S.N. The universal data connection. *IEEE Spec.* July 1987, 31–37.
21. Peach, D.F., Trends Toward a More Stress-Resistant Fiber Optic Telecommunication Installation NTIA Technical Memorandum 86-116, U.S. Department of Commerce, August 1986.
22. Pigliacampi, J.J., and Riewald, P.G., Performance Bases for the Use of Aramid Fiber in Marine Applications IEEE Technical Paper, CH1478, 440–449, 1979.
23. Poehlmann, K.M., Meeting Future C31 Needs with Fiber Optics, the RAND Corporation, P-7093-RGI, May 1985.
24. Porcello, L.J. Optical Processing Operations in Synthetic Aperture Radar Systems, SPIE. (*Effect. Util. of Opt. in Radar Syst.*) 1977, 128, 108–117.
25. Researchers foresee sharp increase in military photonics applications. *Aviation Week and Space Technology*, Fink, Donald, 30 January 1989, cd, 56–57.
26. Sanferrare, R.J. Terrestrial Lightwave systems. *AT&T Tech. J.* 1987, 66, 95–107.

27. Schwartz, M., *Information Transmission, Modulation, and Noise, A Unified Approach to Communication Systems*, 2d ed., McGraw-Hill Book Company, New York, 1970.
28. Shannon, C.E., and Weaver, W., *The Mathematical Theory of Communication*, University of Illinois Press, Urbana, IL, 1949.
29. Shapero, D.C. (ed.), *Photonics: Maintaining Competitiveness in the Information Era, Panel on Photonics Science and Technology Assessment*, National Research Council, National Academy Press, Washington, D.C., 1988.
30. Sherrets, L., and Digest, L., *Shuford, Richard S., an Introduction to Fiber Optics, Part 2: Connections and Networks, BYTE*, January 1985, 1, 197–207, 1988.
31. Shuford, R.S. An introduction to fiber optics, Part 1. *Byte* December 1984, 121.
32. Sigel, G.H., Jr. Fiber transmission losses in high-radiation fields. *Proc. IEEE* 1980, 68, 1236–1240.
33. Vernon, J. Military Taps Into optical fiber. *Def. Electron.* June 1987.

9 Fiber Optic Sensors for Harsh Environment Applications

9.1 INTRODUCTION

The various advantages of fiber optic sensing techniques make it attractive to many industrial sensor applications. They are small in size, passive, immune to electromagnetic interference, and, most importantly, they are resistant to harsh environments and have the ability to perform the sensor distributed very effectively. In addition, its wired and wireless assets make it able to easily integrate into large-scale optical communication networks and systems. [1].

9.2 GRATING TYPES FOR HARSH ENVIRONMENTAL SENSING

Fiber Bragg gratings (FBGs) are found in several different types in terms of manufacturing, installation, and application. Recently, a new classification mechanism has been proposed that relies primarily on the formation mechanism [1–4]. Accordingly, these labels will be adopted during the explanation and as detailed.

9.2.1 TYPE I GRATINGS

This type, classified as the standard fiber gratings, is formed from germane combined with silicate fiber, small refractive index changes; a single UV photon excites oxygen deficiency defect centers with absorption bands around 244 nm [1, 5]. For large refractive index changes, defect formation is accompanied by densification of the glass matrix [1, 6]. This grating type exhibits a positive change with negative temperature dependence [1, 7].

The negative behavior with heat is the result of the thermal properties of the raised excited cells formed during the formation of the grating. With high temperatures, energy is already absorbed by the carriers into shallow traps that are sufficient to escape and return to the ground state. Thus, the remaining carriers are connected, resulting in a more stable state and can also relax into the ground state if the temperature rises further [1].

Physically, most unstable terminals suffer from degradation at low temperatures, so that until this point is overcome, they are typically pulled at temperatures higher than the operating temperature designed for long-term stability. In practice, this type is considered inappropriate in terms of its applicability to harsh environments where temperatures are very high (i.e. temperatures > 450°C), as most of the refractive index change is annealed out at these temperatures [1].

9.2.2 Type II Gratings

Using high-peak-power pulsed ultraviolet laser sources, such as krypton fluoride–based excimer lasers, high reflectivity gratings have been inscribed with a single laser pulse [1, 8]. Gratings with high refractive index can obtained by using a high-peak-power pulsed UV light, such as krypton fluoride–based excimer lasers [1, 8]. This type is the result of a multi-stage ionization process dependent on the threshold, similar to the damage caused by laser in bulk optics. Therefore, they are often referred to as "damage." This type is characterized by being stable at higher temperatures, above 1000°C, so it is used to manufacture grating arrays while fibers are pulled on the draw tower [1, 9].

Single-shot exposure tends to produce grating structures that can suffer from a significant scattering loss. This process also has a tendency to reduce the reliability and mechanical strength of fibers [1].

9.2.3 Type In (Formerly Type IIA)

During the process of grating writing by UV- induced light in highly stressed Ge-doped core fibers, it has been observed that, under certain conditions, Type I grating grows and then decreases in strength as it saturates. Continuous exposure to UV light leads to a secondary grating growth appearing in a steady blue shift in the spectral response, with a negative change in the refractive index [1, 10]. Under optimum fabrication conditions, Type In grating structures are stable up to 700°C [1, 11]. In general, an agreement based on the mechanism associated with the change of the final index involves some kind of stress relief within fiber [1].

9.2.4 Chemical Composition/Regenerated Gratings

The required photosensitivity for Type I grating structures can be obtained through a fabrication process called hydrogen loading. In this process, the fiber is exposed to high-pressure hydrogen gas at room temperature [1, 12]. When the gas permeates the glass matrix, the glass is fully saturated and the fibers are loaded, causing ultraviolet irradiation to break down the hydrogen and form defects in its structures. This increases the levels of measurable indicator change [1, 12].

After writing by ultraviolet radiation and the release of non-reacting hydrogen from the fiber core, the precise splicing of the meshes at a medium temperature of 600–700°C leads to the interstitial diffusion of hydroxyl groups to form water molecules within the glass that becomes very thermally stable [1, 13]. In medium temperatures, the solubility leads to the erosion of the primary sieve. More heating at higher temperatures leads to the generation of a new grating structure and longer wave length. The adjustment of the indicator of this new sieve is much lower than the original seedling by about an order of magnitude [1].

However, they are stable at high temperatures and can be recycled frequently to temperatures above 1000°C [1, 14, 15]. A residual indicator is usually less than 10^{-4}, resulting in low reflectivity if it is less than 1 cm in length. Because of its low index modulation, it is less vulnerable to the scattering losses that are often associated with

Type II gratings. More recently, it has been shown that hydrogen should not be present during laser etching but also during thermal regeneration [1, 16].

9.2.5 FEMTOSECOND PULSE DURATION INFRARED LASER-INDUCED GRATINGS

Instead of using UV radiation, ultrahigh energy radiation resulted by femtosecond pulse duration infrared (fs-IR) laser systems obtained large index changes in the detector in large quantities for the manufacture of unsafe waveguides [1, 17]. This mechanism of change varies significantly from this laser pointer to the optical sensitivity mechanism of ultraviolet light that depends on the formation of the color center. This method is thought to be the product of multiple nonlinear absorption/ionization processes, resulting in material pressure and/or malformation based on exposure intensity [1, 17].

Above the strength of the material-based ionization threshold, multiphoton ionization (MPI)-induced dielectric breakdown causes local fusion, pressure of the material, and vacuum formation, which causes a change in the index and is always up to the temperature of the glass transition of the material [1]. The characteristics of the change of the resulting index are similar to the characteristics of Type II gratings, which are caused by the UV light, but with one significant difference [1].

Due to the ultrashort duration of the femtosecond source, the induced dielectric collapse is quickly suppressed when the beam stops, so there is virtually no "damage" outside the irradiated zone [1]. This process can produce grating structures with excellent spectral performance compared to their ultraviolet counterparts. Below the threshold current, another system of induced indicator changes, which can be erased by thermodynamics, was observed at temperatures below the glass temperature [1, 18].

In the regime below the threshold current, a multiphoton absorption is likely to cause malformation, as well as pressure of materials similar to those in the UV-induced Type I index changes in photosensitive Ge-Si glasses. Currently, there are two main approaches to writing Bragg gratings with fs-IR laser sources. In the first method of clarification, a specialized phase mask that was carefully drilled was used to maximize the association of infrared laser radiation incident in ±1 orders [1, 19].

Gratings were written in standard optic fiber and in "non-UV photosensitive" pure silica core fibers [1, 20]. Using the phase mask leads to the generation of a sinusoidal interference field and results in a non-sinusoidal modulated fiber Bragg grating structure due to the nonlinear-induced index change processes [1, 21]. The phase mask method produced both Type I- and Type II-induced index changes in silica-based glass fibers [1, 21].

9.3 HARSH ENVIRONMENTAL SENSING APPLICATIONS OF FBGs

9.3.1 HIGH TEMPERATURE

The advantages that FBGs enjoy as a high thermal resistivity, such as the rinsing or those written with 800 nm femtoseconds, will open up many opportunities in various industrial sectors characterized by harsh environments such as power plants, turbines, combustion, and space [1]. The properties of FBGs written with fs-IR radiation above the threshold current were studied in both the SM-28 and growth factor (GF) of Corning and the pure fiber optic silica (PFOS) nucleus over the long term [1, 22–25].

Bragg gratings with large index modulations (Δn) were inscribed in all fiber types and were obtained by heating them above 1000°C in increments of 100°C and in 1-hour intervals in a special furnace [1]. Generally, FBGs are loosely positioned to avoid external pressure. Then, once the temperature reaches 1000°C, this is maintained for 150 hours, controlled by FBG transmission spectra [26]. An important observation throughout the test period was that there was no deterioration in the grating strength $\Delta n = 1.7 \times 10^{-3}$ [1, 26].

The increase in the Δn is observed by the slight rise in the reflective grating reflectivity, and is the result of two types of change in the Δn that is written at the same time. The peaks of the interference pattern are so complex that they are sufficient to support the glass in the fibers, which means that the change in the amount of the indicator is always correlated with the temperature [1]. Through the interference, the intensity characteristics often is below the Type II threshold; however, some Type I grating index change is generated [1].

Since the fiber gratings are always protected by a cover, the permanent Type II index change remains, and Type I index change is erased. [1] When the temperature of FBG was increased and maintained at 1050°C for 100 hours, Δn decreased slightly from 1.7×10^{-3} to 1.6×10^{-3} [1, 27]. At the end of the experiment, a drift in the Bragg resonance by 0.2 nm was detected. When the fibers are pre-lined at high temperatures to mitigate the remaining stresses, FBGs of Type II fs-IR work up to 1200°C [1, 27].

Reflection is the result of an order of magnitude higher than what can be obtained by using Type II UV or retina. On the other hand, to operate in a high-temperature environment, a very important point must be noted: the challenge of encapsulating the silica fiber optic sensor [1]. At temperatures closer to or above 1000°C in the air, the fibers will lose the only stereotypes made of insulated silica and all its mechanical strength. The fiber itself can survive hundreds of hours at 1000°C [1, 29] when left untouched, after which any fiber treatment after testing is not possible because the fibers become very fragile [1, 29].

It is clear that fiber optics suffer from deterioration in their mechanical properties and are very severe when tested in oxidizing environments at extremely high temperatures. Therefore, protection conditions must be available to avoid loss of their properties, and this can be accomplished by using an appropriate package that avoids high temperature [1]. There are many methods of protection techniques. The most obvious option is to paint on fiber after writing the accompanying material or writing through the protective layer. Metallic coatings are preferred for high-temperature applications, where golden paint is rated at 700°C, but this classification is not suitable for temperatures close to 1000°C [1, 30].

For temperatures above 1200°C, silica-based fiber optics are not suitable for many applications with extreme conditions. Therefore, monocrystalline sapphire fibers are more successful in applications of high-temperature thermodynamics since they contain a glass transition temperature of about 2030°C [1]. Unlike conventional single-mode traditional fiber, the process of manufacturing sapphire fibers is in the form of rods without a layer of cladding. This makes the multicolored sapphire evidence highly sensitive and sensitive to bending loss and pattern conversion [1].

With commercially available fiber diameters, beam propagation within fibers is highly variable, especially in the 1,550 nm communication wavelengths. An example of fibrous diameters is shown by Mihailov [1]. Sapphire fibers are often based on the Fabry–Pérot structures within the fibers that produce the interference signal in broadband but have a problem with temperature changes [1, 31]. These devices are used effectively as point sensors.

9.3.2 HIGH RADIATION

For nuclear applications, conventional measuring devices are not suitable in such environments, where temperatures are very high and are accompanied by chemical contamination and high levels of electromagnetic interference [1]. Neutron bombardment also results in harmful effects on materials, where a thermal output can drift under intense radiation as the two metals that form the thermocouple gradually move to different elements [1, 32].

Fiber Bragg grating sensors, which are manufactured by means of a pattern obtained by ultraviolet radiation, are assumed to be able to withstand radiation in low-flow nuclear environments, even for extended periods of time [1, 33]. But in high-camouflage environments, Ge-doped fiber is vulnerable to attenuation caused by radiation that makes the fibers unclear after a period of time. This formation process is similar to the fault-making mechanism that occurs during UV-type Type I writing in Ge-reinforced fiber. [1].

In order to obtain reliable measurements of stress and temperature, it is necessary to insert grid barriers in radiation-resistant fibers and resistance to radiation-induced attenuation (RIA), such as fibers containing pure or dehydrated silica nuclei [1, 34]. Because of the design characteristics of these fibers, which are inherently non-light-sensitive, standard codification techniques are ineffective. Therefore, the clamps are easily inserted into high-radiation-enhanced fiber [1, 35] using the fs-IR method with a phase mask in the first and second genres [1, 36]. In these species, a very slight change in the spectral quality of the network was observed even after taking a dose of radiation of 100 kGy [1, 31].

9.3.3 MULTI-PARAMETER SENSING IN HARSH ENVIRONMENTS

For FBG sensors, one of the most important sensor parameters to be used to determine a target is the Bragg wavelength of the network. However, it is often difficult to distinguish between different effects, such as temperature and pressure, because all the different effects may overlap at the same time, making the selection process complex [1]. To overcome this challenge, a different Bragg pan is often used for each of the parameters involved in each case, but, unfortunately, this procedure may result in a very complex sensory composition [1, 37].

To avoid multiple fibers for sensing multiple parameters, a variety of alternatives were used to avoid this complexity [1, 38-40], FBGs in different diameters [1, 42], the use of second-degree diffraction from the first saturation of 1,550 nm grating [1, 43], FBGs in fiber birefringent [1, 44] of the roads. These techniques have been developed mainly for the special case of stress and temperature discrimination [1, 45].

9.3.4 HIGH-PRESSURE SENSING

When subjected to very high pressure, the first types of FBG undergo negative waveforms that are proportional to the level of hydrostatic pressure [46]. This shift is somewhat modest at about 3 pm/MPa, but it is observed that it changes linearly at pressures of up to 10,000 psi (70 MPa). However, the sensitivity of the nets can be increased to approximately 20-fold pressure by using synthetic rubber paint and polyurethane paint [1, 47]. Recently, its sensitivity has been enhanced by a commitment to a composite carbon fiber structure to FBG that has become three times the size and has been introduced at pressures of up to 70 MPa [1, 48].

In these cases, pressure measurements are made at or near room temperature. In the event of sudden changes in the measured environment leading to both high pressures and high temperatures simultaneously, the multi-parameter pressure/heat sensor based on Type I grating structures is not suitable for the necessary measurements [1]. A fiber-based geometry technique was used to perform simultaneous measurements of high pressure and temperature [1, 49]. It was observed that with increased pressure inside the lateral holes there was a decrease in the separation of polarization-dependent responses [1].

The changes in the temperature were observed by the shift in the wavelength corresponding to each resonance [1]. Using technology, the fs-IR Type II is inserted in order to produce a sensor, so that it is able to observe changes that may result in temperature and pressure at the same time in harsh environments [1], 50, 51].

These sensors have been proven to work in harsh environments and have been found to be able to measure at pressures ranging from 15 to 2,400 psi (0.1 to 16.5 mpg) at temperatures up to 800°C [1]. In addition, the core area of the fibers has been identified so that the guided position has some conjugation in the area with the aperture. It can also be used to measure the refractive index of materials or liquids within the lateral hole [1, 52].

9.3.5 HIGH-PRESSURE HYDROGEN DETECTION

Recently, there has been growing interest in the use of hydrogen in multiple applications, particularly in energy production and alternative fuels in the automotive sector. [1]. Hydrogen itself should not generate any contaminants during the combustion process, adding to its reproducibility from renewable sources. Hydrogen fuel cells are expected to be among the most promising industrial applications, but hydrogen is also being studied for petrochemical operations and other uses [1, 53].

Hydrogen is usually stored under high pressure. In the nuclear industry, next-generation reactors such as supercritical-water cooled reactors (SCWR) and very-high-temperature reactors (VHTR) are seen as effective ways of producing large amounts of hydrogen, but high pressures and temperatures are needed in these processes [1], approximately up to 25 MPa, 1000°C. Therefore, the need for high-precision sensors to detect hydrogen must be available in such conditions [1, 54]. These sensors should be environmentally friendly in terms of the nature of narrow spaces, so they are small, powerful, inexpensive, reliable, recoverable with long-term stability, and are capable of continuing to work in explosive environments [1].

These sensors were tested to detect hydrogen in the concentration range of 1 to 4% [1]. These sensors can be observed by observing the spectral shift of the Bragg wavelength caused by the stress caused by changes in the size of the palladium layer [55] or by changes in the palladium refractive index [1, 56].

One way to increase the sensor response time is to heat the fiber. Because it is highly sensitive to small concentrations of hydrogen, the palladium system of hydrogen depends mainly on hydrogen pressure and temperature, which can be in two amorphous phases called α and β, separated by phase transfer [1]. One must turn to an important point, namely, the operation of the sensor in the field of pressure temperature in either of these stages is considered a very urgent necessity [1, 57].

9.3.6 HIGH RELIABILITY FBGs FOR HIGH-STRAIN MEASUREMENTS

The high capacity in which the FBG sensor scales to measure high voltage levels makes it very suitable for monitoring structural integrity [1]. The standard method of UV lithography requires the removal of a protective polymer cover that is highly absorbed into ultraviolet light prior to, and replaced after, FBG manufacture. Coating and paint removal processes take a lot of time and also threaten the mechanical safety of fibers, thus reducing the maximum value of the measurable strain [1, 58].

By using the fs-IR laser and the phase mask technique, the gratings were successfully inscribed through the acrylate coating of standard single-mode fiber (SMF)-28 fiber [1, 59] and high numerical aperture (NA) fiber [1, 60], after the fiber's photosensitivity to fs-IR radiation was enhanced by hydrogen loading [1, 61]. Index modulations of up to 1.4×10^{-3} were induced in the high NA fibers, with strengths remaining at 75% to 85% of the pristine fiber value [1].

For stress-sensing applications, FBG is preferred in non-contact fibers to bend polyamide, as the high-polymer layer acts as a conduit for better transmission of mechanical information to fibers compared to thick acrylic layers and smoothness [1]. Using the same method, but in highly non-polyamide-sensitized NA fibers, 1×10^{-4} indexing configurations were generated, with residual strengths at ~50% of the original fiber value [1, 62].

9.4 SUMMARY

This chapter reviewed the use of optical fiber Bragg grating sensors for harsh environment applications (i.e. high temperature, high radiation, high pressure sensing, high pressure hydrogen detection, high reliability FBGs for high strain measurements),and the grating types for harsh environmental sensing (i.e. chemical composition/regenerated gratings, femtosecond pulse duration infrared laser induced gratings).

REFERENCES

1. Mihailov, S.J. Fiber Bragg grating sensors for harsh environments. *Sensors (Basel)* 2012, 12, 1898–1918.
2. Hill, K.O., Malo, B., Bilodeau, F., Johnson, D.C., and Albert, J. Bragg gratings fabricated in monomode photosensitive optical fiber by UV exposure through a phase mask. *Appl. Phys. Lett.* 1993, 62, 1035–1037.
3. Kersey, A.D., Davis, M.A., Patrick, H.J., LeBlanc, M., Koo, K.P., Askins, C.G., Putnam, M.A., and Friebele, E.J. Fiber grating sensors. *J. Lightwave Technol.* 1997, 15, 1442–1463.
4. Canning, J. Fibre gratings and devices for sensors and lasers. *Laser & Photon. Rev.* 2008, 2, 275–289.
5. Meltz, G., Morey, W.W., and Glenn, W.H. Formation of Bragg gratings in optical fibers by a transverse holographic method. *Opt. Lett.* 1989, 14, 823–825.
6. Poumellec, B., Niay, P., Douay, M., and Bayon, J.F. U.V. Induced densification during Bragg grating writing. In: Proceedings of OSA Conference on Photosensitivity and Quadratic Nonlinearity in Glass Waveguides, Portland, OR, USA, 9–11 September 1995.
7. Erdogan, T., Mizrahi, V., Lemaire, P.J., and Monroe, D. Decay of ultraviolet-induced fiber Bragg gratings. *J. Appl. Phys.* 1994, 76, 73–80.
8. Archambault, J.-L., Reekie, L., and Russell, P.J. High reflectivity and narrow bandwidth fibre gratings written by single excimer pulse. *Electron. Lett.* 1993, 29, 28–29.
9. Askins, C.G., Putman, M.A., Williams, G.M., and Friebele, E.J. Stepped wavelength optical fiber Bragg grating arrays fabricated in line on a draw tower. *Opt. Lett.* 1994, 19, 147.
10. Xie, W.X., Niay, P., Bernage, P., Douay, M., Bayon, J.F., Georges, T., Monerie, M., and Poumellec, B. Experimental evidence of two types of photorefrecitve effects occurring during photoinscriptions of Bragg gratings within germanosilicate fibres. *Opt. Commun.* 1993, 104, 185–195.
11. Groothoff, N., and Canning, J. Enhanced type IIA gratings for high-temperature operation. *Opt. Lett.* 2004, 29, 2360–2362.
12. Lemaire, P.J., Atkins, R.M., Mizrahi, V., and Reed, W.A. High pressure H2 loading as a technique for achieving ultrahigh UV photosensitivity and thermal sensitivity in GeO2 doped optical fibres. *Electron. Lett.* 1993, 29, 1191–1193.
13. Fokine, M. Underlying mechanisms, applications, and limitations of chemical composition gratings in silica based fibers. *J. Noncrystal. Solids* 2004, 349, 98–104.
14. Zhang, B., and Kahrizi, M. High-temperature resistance fiber Bragg grating temperature sensor fabrication. *IEEE Sensors J.* 2007, 7, 586–591.
15. Canning, J., Bandyopadhyay, S., Stevenson, M., Cook, K., and Bragg, F. Grating sensor for high temperature application. In: Proceedings of Joint Conference of Opto-Electronics and Communications Conference and Australian, Conference on Optical Fibre Technology, Sydney, Australia, 7–10 July 2008.
16. Lindner, E., Canning, J., Chojetzki, C., Brückner, S., Becker, M., Rothhardt, M., and Bartelt, H. Post-hydrogen-loaded draw tower fiber Bragg gratings and their thermal regeneration. *Appl. Opt.* 2011, 50, 2519–2522.
17. Davis, K.M., Miura, K., Sugimoto, N., and Hirao, K. Writing waveguides in glass with a femtosecond laser. *Opt. Lett.* 1996, 21, 1729–1731.
18. Sudrie, L., Franco, M., Prade, B., and Mysyrowicz, A. Study of damage in fused silica induced by ultra-short IR laser pulses. *Opt. Commun.* 2001, 191, 333–339.
19. Mihailov, S.J., Smelser, C.W., Lu, P., Walker, R.B., Grobnic, D., Ding, H., Henderson, G., and Unruh, J. Fiber Bragg gratings made with a phase mask and 800-nm femtosecond radiation. *Opt. Lett.* 2003, 28, 995–997.

20. Mihailov, S.J., Smelser, C.W., Grobnic, D., Walker, R.B., Lu, P., Ding, H., and Unruh, J. Bragg gratings written in All-SiO2 and Ge-doped core fibers with 800-nm femtosecond radiation and a phase mask. *J. Lightwave Technol.* 2004, 22, 94–100.

21. Smelser, C.W., Mihailov, S.J., and Grobnic, D. Formation of type I-IR and type II-IR gratings with an ultrafast IR laser and a phase mask. *Opt. Express* 2005, 13, 5377–5386.

22. Martinez, A., Dubov, M., Khrushchev, I., and Bennion, I. Direct writing of fibre Bragg gratings by a femtosecond laser. *Electron. Lett.* 2004, 40, 1170–1172.

23. Martinez, A., Dubov, M., Khrushchev, I., and Bennion, I. Photoinduced modifications in fiber gratings inscribed directly by infrared femtosecond irradiation. *IEEE Photon. Technol. Lett.* 2006, 18, 2266–2268.

24. Martinez, A., Khrushchev, I.Y., and Bennion, I. Thermal properties of fibre Bragg gratings inscribed point-by-point by infrared femtosecond laser. *Electron. Lett.* 2005, 41, 176–178.

25. Mihailov, S.J., Grobnic, D., Smelser, C.W., Lu, P., Walker, R.B., and Ding, H. Bragg grating inscription in various optical fibers with femtosecond infrared lasers and a phase mask. *Opt. Mater. Express* 2011, 1, 754–765.

26. Grobnic, D., Smelser, C.W., Mihailov, S.J., and Walker, R.B. Long-term thermal stability tests at 1000°C of silica fibre Bragg gratings made with ultrafast laser radiation. *Meas. Sci. Technol.* 2006, 17, 1009–1013.

27. Li, Y., Yang, M., Wang, D.N., Lu, J., Sun, T., and Grattan, K.T. Fiber Bragg gratings with enhanced thermal stability by residual stress relaxation. *Opt. Exp.* 2009, 17, 19785–19790.

28. Méndez, A., and Morse, T.F. *Specialty Optical Fibers Handbook*, Elsevier Academic Press, San Diego, CA, USA, 284, 2007.

29. Grobnic, D., Mihailov, S.J., Walker, R.B., and Smelser, C.W. Self-packaged Type II femtosecond IR laser induced fiber Bragg grating for temperature applications up to 1000°C. *Proc. SPIE* 2011, 7753, doi:10.1117/12.886396.

30. Putnam, M.A., Bailey, T.J., Miller, M.B., Sullivan, J.M., Fernald, M.R., Davis, M.A., and Wright, C.J. Method and apparatus for forming a tube-encased Bragg grating. US Patent 6,298,184, 2001.

31. Wang, A., Gollapudi, S., May, R.G., Murphy, K.A., and Claus, R.O. Sapphire optical fiber-based interferometer for high temperature environmental applications. *Smart Mater. Struct.* 1995, 4, 147–151.

32. Grobnic, D., Mihailov, S.J., Smelser, C.W., and Ding, H. Sapphire fiber Bragg grating sensor made using femtosecond laser radiation for ultrahigh temperature applications. *IEEE Photon. Technol. Lett.* 2004, 16, 2505–2507.

33. Fernandez, A.F., Gusarov, A., Brichard, B., Decréton, M., Berghmans, F., Mégret, P., and Delchambre, A. Long-term radiation effects on fibre Bragg grating temperature sensors in a low flux nuclear reactor. *Meas. Sci. Technol.* 2004, 15, 1506–1511.

34. Wijnands, T., De Jonge, L.K., Kuhnhenn, J., Hoeffgen, S.K., and Weinand, U. Optical absorption in commercial single mode optical fibers in a high energy physics radiation field. *IEEE Trans. Nucl. Sci.* 2008, 55, 2216–2222.

35. Aikawa, K., Izoe, K., Shamoto, N., Kudoh, M., and Tsumanuma, T. Radiation-resistant single-mode optical fibers. *Fujikura tech. rev.* 2008, 37, 9–13.

36. Grobnic, D., Henschel, H., Hoeffgen, S.K., Kuhnhenn, J., Mihailov, S.J., and Weinand, U. Radiation sensitivity of Bragg gratings written with femtosecond IR lasers. *Proc. SPIE* 2009, 7316, 73160C.

37. Reekie, L., Dakin, J.P., Archambault, J.-L., and Xu, M.G. Discrimination between strain and temperature effects using dual-wavelength fibre grating sensors. *Electron. Lett.* 1994, 30, 1085–1087.

38. James, S.W., Dockney, M.L., and Tatam, R.P. Simultaneous independent temperature and strain measurement using infibre Bragg grating sensors. *Electron. Lett.* 1996, 32, 1133–1134.

39. Echevarria, J., Quintela, A., Jauregui, C., and Lopez-Higuera, J.M. Uniform fiber Bragg grating firstand second-order diffraction wavelength experimental characterization for strain-temperature discrimination. *IEEE Photon. Technol. Lett.* 2001, 13, 696–698.

40. Urbanczyk, W., Chmielewska, E., and Bock, W.J. Measurements of temperature and strain sensitivities of a two-mode Bragg grating imprinted in a bow-tie fibre. *Meas. Sci. Technol.* 2001, 12, 800–804.

41. Frazão, O., Ferreira, L.A., Araújo, F.M., and Santos, J.L. Applications of fiber optic grating technology to multi-parameter measurement. *Fiber Integr. Opt.* 2005, 24, 227–244.

42. Gower, M.C., and Mihailov, S.J. Recording of efficient high-order Bragg reflectors in optical fibres by mask image projection and single pulse exposure with an excimer laser. *Electron. Lett.* 1994, 30, 707–709.

43. Grobnic, D., Mihailov, S.J., Smelser, C.W., and Walker, R.B. Multiparameter sensor based on single high-order fiber Bragg grating made with IR-Femtosecond radiation in single-mode fibers. *IEEE Sensors J.* 2008, 8, 1223–1228.

44. Chan, C.F., Chen, C., Jafari, A., Laronche, A., Thomson, D.J., and Albert, J. Optical fiber refractometer using narrowband cladding-mode resonance shifts. *Appl. Opt.* 2007, 46, 1142–1149.

45. Thomas, J., Jovanovic, N., Becker, R.G., Marshall, G.D., Withford, M.J., Tünnermann, A., Nolte, S., and Steel, M.J. Cladding mode coupling in highly localized fiber Bragg gratings: Modal properties and transmission spectra. *Opt. Exp.* 2011, 19, 325–341.

46. Jewart, C.M., Wang, Q., Canning, J., Grobnic, D., Mihailov, S.J., and Chen, K.P. Ultrafast femtosecond-laser-induced fiber Bragg gratings in air-hole microstructured fibers for hightemperature pressure sensing. *Opt. Lett.* 2010, 35, 1443–1445.

47. Chen, T., Chen, R., Jewart, C., Zhang, B., Cook, K., Canning, J., and Chen, K.P. Regenerated gratings in air-hole microstructured fibers for high-temperature pressure sensing. *Opt. Lett.* 2011, 36, 3542–3544.

48. Maier, R.R.J., Jones, B.J.S., Barton, J.S., McCulloch, S., Allsop, T., Jones, J.D.C., and Bennion, I. Fibre optics in palladium-based hydrogen-sensing. *J. Opt. A: Pure Appl. Opt.* 2007, 9, S45–S59.

49. Zalvidea, D., Díez, A., Cruz, J.L., and Andrés, M.V. Hydrogen sensor based on a palladium-coated fibre-taper with improved time-response. *Sens. Actuat. B Chem.* 2006, 114, 268–274.

50. Trouillet, A., Marin, E., and Veillas, C. Fibre gratings for hydrogen sensing. *Meas. Sci. Technol.* 2006, 17, 1124–1128.

51. Malo, B., Albert, J., Hill, K.O., Bilodeau, F., and Johnson, D.C. Effective index drift from molecular hydrogen diffusion in hydrogen loaded optical fibres and its effect on Bragg grating fabrication. *Electron. Lett.* 1994, 30, 442–444.

52. Geernaert, T., Becker, M., Mergo, P., Nasilowski, T., Wojcik, J., Urbanczyk, W., Rothhardt, M., Chojetzki, C., Bartelt, H., Terryn, H., Berghmans, F., and Thienpont, H. Bragg grating inscription in GeO2 doped microstructured optical fibers. *J. Lightwave Technol.* 2010, 28, 1459–1467.

53. Grobnic, D., Mihailov, S.J., Walker, R.B., Cuglietta, G., and Smelser, C.W. Hydrogen detection in high pressure gas mixtures using a twin hole fibre Bragg grating. *Proc. SPIE* 2011, 7753, 77537D-1–77537D-4.

54. Mihailov, S.J., Grobnic, D., Ding, H., Smelser, C.W., and Broeng, J. Femtosecond IR laser fabrication of Bragg gratings in photonic crystal fibers and tapers. *IEEE Photon. Technol. Lett.* 2006, 18, 1837–1839.

55. Zalvidea, D., Díez, A., Cruz, J.L., and Andrés, M.V. Hydrogen sensor based on a palladium-coated fibre-taper with improved time-response. *Sens. Actuat. B Chem.* 2006, 114, 268–274.
56. Trouillet, A., Marin, E., and Veillas, C. Fibre gratings for hydrogen sensing. *Meas. Sci. Technol.* 2006, 17, 1124–1128.
57. Mihailov, S.J., Grobnic, D., and Smelser, C.W. Efficient grating writing through fibre coating with femtosecond IR radiation and phase mask. *Electron. Lett.* 2007, 43, 442–443.

...

10 Other Applications

10.1 INTRODUCTION

In recent years, fiber optic technology has had a great deal of control over applications and in various fields of industry due to its many advantages of data transmission speed, immunity to electromagnetic waves, and their simplicity and ease of installation compared to conventional technologies. These properties made it very attractive for sensor and control applications and tools. In these areas, optical fiber has a significant impact, making it a fierce competitor [1–3].

10.2 SHIP CARGO HANDLING SYSTEMS

Sea transport techniques are under pressure to reduce costs and improve efficiency. To achieve this, fiber optic technology has an important role to play in the design of modern commercial ships [4]. The new requirements for modern vessels impose greater use of electrical appliances, automation, and control to increase circulation and reduce crew to work [4, 5].

Fiber optic technology has tremendous potential by using a single-fiber optic string as a sensor to measure many parameters as some electrical and non-electrical values in ship systems [6]. One of the most important advantages of optical fiber technology compared to conventional technology is its immunity against any electrical interference and its ability to protect against corrosion [6]. Moreover, this type does not necessarily need any electrical path, since in its simplest form, the fiber optic sensor consists of a light source and an optical fiber as a data transmission channel, a sensing element, and a detector [6].

The light source may be a broadband, a light-emitting diode (LED), or a laser diode (LD), depending on the nature of the sensing parameter. The type of sensor when the sensing element is an essential part of the fiber is called an intrinsic sensor [6]; if the fiber is only used for data transporting from the sensing element, the arrangement is known as an extrinsic sensor [6].

Physical properties of light such as amplitude, frequency, phase, polarization and the sensing element may be affected by the change in the environment surrounding the sensing part of the sensor. These changes can easily be detected and recorded by optical measuring instruments [1].

The optical fiber works double—as a sensor as well as a channel to transmit optical sensor signals. The state of the vessel system is monitored by fiber optic sensor technology by a set of central control stations located along the vessel bridge via the optical cable. Critical data can be transferred from the central station via the satellite link to the party involved in the maritime operation [2]. In addition, it is possible for a person to take charge of monitoring the vessel from a fiber optic network to access the monitoring stations via the Internet and to obtain real-time access to the data necessary for the specific ship system and compare it with a computer model to solve the problem [1].

10.2.1 Longitudinal/Transversal Ship Hull Strength Monitoring

Strain is an element that shows the internal or compressive tension of the structure of the vessel resulting from the difference between local loading forces and buoyancy. The longitudinal force in general is represented by the maximum bending moment that can be sustained by the ship's cross section [7]. The structure of the vessel shall be divided, in the longitudinal direction, into sections for the protection of the vessel and its crew and the cargo it carries from potential serious damage resulting from accidents [8].

National and international rules and regulations have been established and are the minimum conditions for the ship's power. When a vessel floats or moves through water, its structure is exposed to different types of forces, so all ships' points must be able to withstand these forces; they are primarily designed in the shape of the ship's structure [8].

There continues to be development in order to find a new method to determine the strength of the ship and its body for the purpose of predicting its behavior under different environmental conditions [1, 9]. The important point in ship design is always how to determine the strength of the ship relative to its body, and to find its points. The characteristics of optical sensing technology can be used to achieve this by checking the cation of transverse, longitudinal, and local forces for temporary loading and to detect any defects or distortions that may occur and to assist in making timely decisions in the event of an emergency arising from an accident [9].

One of the most important challenges facing ships is the problem of erosion due to their working environment, as the external structure is always in the water line as well as the upper brightness, which is subject to external conditions. The structures of the interior reservoirs are also the areas most prone to corrosion [1]. There are many techniques that can reduce the amount of corrosion in the ship, but most of them may lead to reduced body thickness. To ensure their validity, all ships are subject to periodic inspection by the Classification Society. In this case, the case data of the vessel structure element obtained by the distributed fiber sensor is useful for the rating community to verify the longitudinal force [1, 9].

One of the most important factors affecting longitudinal force is the bottom of the ship's structure and surface. Therefore, the distribution of fiber optic module points is as shown by Cardis [9]. Figure 2 in Cardis's book [9] represents a graphical chart of the vessel section. This form is suitable for a double-hull carrier, bulk conveyor, container carrier, single-hull conveyor, or any other type of vessel used for commercial shipping operations [9].

This system is very useful for monitoring the health condition of the structure of the vessel and to predict the amount of damage that it can be exposed to, especially when loading it to the maximum limit [1]. This system can assist the ship captain by providing him with sufficient information on the amount and tolerability of the longitudinal force [1]. Also, the distributed sensing system along the hull can contribute not only to erosion control but can be used to improve ship design and operational availability [1].

Many data can be collected under various marine conditions by the ship's captain or by others interested in their activities, such as ship owner, classification societies, and insurance companies, in order to determine the actual status of the ship and determine the nature of the necessary fees to be taken [1].

10.2.2 MEASURING THE MASS LOADED OR DISCHARGED CARGO

The method of comprehensive surveys was adopted as an effective, accurate, and appropriate method for determining weight on board ships, where it is commercially agreed that the weight of cargo transported by sea should be determined by the survey project. This project is a way to determine the weight of the mass loaded on the vessel by measuring the draft measurements on drafts of the marks in the bow, center, and rear [10].

Using this method, the accuracy of the load loaded or unloaded may vary due to unexpected and dangerous errors [11]. One of the most important factors that contributes to the determination of weight is the initial and final readings. The accuracy of the calculation can be increased by reducing the percentage of errors by analyzing the readings that have an impact on the draft quarter. The authors concluded that the impact of the error in the reading projects was the most important, especially for the readings of the midship's project [11].

To avoid these errors using classical methods, it was suggested that fiber optic sensors be used instead [1]. These sensors are distributed inside the ship's structure so that the pipelines are connected through the lower valve at the sea level, and the associated sound pipes should be placed on the main pipeline on the inside of the ship's structure against the reading marks on the vessel's outer plate [1].

In addition, the fiber optic sensor must be liquid level to measure sea level measurements built for the internal draft reading system. Sensors should also consist of a plastic fiber cable and an optical time-domain reflectometer (OTDR) device, consisting of a light source for a scanner, an optical detector, a processor, and a display.

10.2.3 LIQUID LEVEL READINGS IN DIFFERENT TANKS

Classic methods commonly used in the determination of fluid quantities in ballast and reservoirs of consumables (fuel oil, fresh water, etc.) and cargo tanks [1] are usually used. So the quantities are presented in special tables and must be approved by an appropriate classification community. The tanks should be operated by a steel gauge bar, covered with a liquid-fixer. The fluid level of sound for the cut and the list should be corrected. The fluid level reading represents the key data entry to the approved tables [1].

The accuracy of taking a liquid level readings method could vary due to both the systematic and the random errors [1]. The liquid level reading can be performed by the use of fiber optic sensors. Liquid level fiber optic sensors, which consist of a coil of fiber built on a cylindrical tube, should have to be vertically positioned on each sounding pipe of the tank. Sensors of the proposed optical liquid level reading system operate on the same principles as sensors explained on the inner draft reading system, and reading the fluid level is the key data entry to the approved tables [1].

The proposed optical liquid level reading system on board a ship should enable more accuracy for the determination of the liquid quantity in ship's tanks (ballast, consumables, and cargo tanks). Furthermore, human error, as one of the most significant error in the tank sounding method, should have to be eliminated by the use of the proposed optical fiber technology [1].

10.2.4 Monitoring Temperature/Humidity in a Ship's Cargo

Ships' holds are a closed place where temperature and humidity are likely to differ significantly between different ships. This change in ambient conditions may have a significant impact on some types of cargo. Alternatively, another type of load may adapt to conditions or create a completely different environment. They therefore do not require continuous testing, but some types of cargo are not very tolerant and may require constant atmospheric monitoring during the vessel's voyage [1].

For multi-purpose ships and cargo, atmospheric data is collected periodically by the ship's administrator. Most ships are equipped with reading equipment for air conditions at each point of shipment and in refrigerated containers [1].

Monitoring climate in shipping technology is important, as it contributes effectively to reducing the heavy losses that may occur. Through monitoring, producers can deliver perishable goods that are thousands of miles away without any significant loss [1]. Through the use of cooling systems, refrigeration companies have managed to maintain temperature/humidity stability with great accuracy, thus extending the shelf life of perishable goods and thus expanding the types of materials that can be shipped in the refrigerated area without damage [1].

Due to the characteristics of fiber optic sensors, they are able to monitor temperature and humidity permanently [1]. Thus, they have the ability to continuously evaluate some important parameters of the load state in the cargo area or in a refrigerated container in each area covered by a suitable fiber optic sensor [12]. The proposed system allows continuous monitoring not only of temperature and humidity but also of the ability to determine differences between them along the optical fiber. The nature of these sensors allows easy placement, which contributes to reliable measurements of the relevant parameters at different test points. The sensor cable measures the parameters at each point along the cable [1, 12].

10.2.5 Load Forces and Strain Effects

At the same time, the length of the large container tanker exceeds 300 meters, and the large-scale force of the formation takes its structure in the longitudinal/transverse direction. In recent years, the speed of large container vessels has increased steadily to maintain the scheduled vessel turnover [1]. In accordance with the international standards of navigation, minimum rules and regulations are defined which assume that the ship is capable of functioning in a healthy manner [12].

With a review of some marine accidents, we conclude that the predictions of marine conditions obtained by traditional vessel measurement methods may not be accurate enough to assess the true and healthy state of the vessel. The climatic conditions of the sea are normally taken by the officer on watch at the navigating bridge [1, 10].

Modern operational requirements for large-scale container vessels require automatic monitoring of the state of the sea through the use of wave radars, which have become an effective tool [1]. These radars help navigation by providing the following important information: wave direction, maximum wave height, wave period, and significant wave height [1].

An important concept in shipping is what is known as container insurance, as it is known through high skin standards. These standards are based on compliance with the Container Insurance Manual, which is mandatory for each vessel, and properly held leather equipment [1]. These measures are one of the basic controls that every container ship must meet to ensure safe voyage and avoid heavy container damage. However, this system is usually affected by load forces due to load weight and stress effects due to vessel movement [1].

10.3 FIBER OPTIC SENSING FOR SATELLITE COMMUNICATIONS

Continuous efforts and research for mass reduction: the traditional wired point-to-point heat sensor, which harnessed the onboard communications satellite system, introduced the hybrid transmission system (HSB) as a future component of the spacecraft. The HSB system relies on remote sensing units as well as fiber optic sensors, both of which can be sequentially connected to the bus structure [13, 14].

HSB is a standard measurement system used in many applications, thanks to its capabilities. However, the focus is on the introduction of fiber optics, especially FBG temperature sensors, as a devastating innovation for the company's satellite platforms [13]. Lightweight fiber optic sensors are directly engraved in mechanically loaded and radiation-tolerant fibers, reducing the need for fiber optic connectors and connections to a minimum. Wherever they are used, the fibers are implanted with a heat adapter similar to the satellite structure or sub-system [13].

The HSB system has been developed under a co-financing scheme from Europe and Germany aimed at satellite communication platforms with long operating periods of more than 15 years within fixed orbits. The company confirms that there is significant compatibility in terms of performance with existing platforms while it is designed with future applications with increased exposure to radiation already considered [13].

The basic installation of the HSB consists of four modules: the power supply unit, the first control unit, the control unit of the interrogator, and the analog front end of the optical fiber interrogation [13]. These units work together with the network of electrical sensors and fiber optics for data processing and signal identification [13].

10.4 SUMMARY

This chapter reviewed the use of optical fiber Bragg gratings in the cargo handling systems of ships (i.e. the longitudinal/transversal ships' hull strength monitoring, measuring mass loaded or discharged cargo, the liquid level readings in different tanks, monitoring temperature/humidity in a ship's cargo, the load forces and strain effects), and in satellite communication.

REFERENCES

1. Ivče, R., Jurdana, I., and Kos, S. Ship's cargo handling system with the optical fiber sensor technology application. *Sci. J. Marit. Res.* 2014, 28, 118–127.
2. Hisham, H.K. *Polymer Optical Fiber Bragg Grating Technology: Spectral Response and Tuning Characteristics*, Lambert Academic Publishing, 2016.
3. Hisham, H.K. *External Cavity Semiconductor Laser Source Based Fiber Bragg Grating for Dense Wavelength Division Multiplexing Systems*, Lambert Academic Publishing, 2015.
4. Dible, J., and Mitchell, P. *Draught Surveys*, MID C Consultancy, 2005.
5. Glavan, B. *Ekonomika morskog brodarstva*, Školska knjiga, Zagreb, 1992.
6. Dakin, J.P., and Brown, R. *Handbook of Optoelectronics*, Vol. I & II. CRC Press, Boca Raton, FL, 2006.
7. Uršić, J. *Čvrstoća broda*, Fakultet strojarstva i brodogradnje Zagreb, Zagreb, 1992.
8. Ivče, R., Jurdana, I., and Mrak, Z. Longitudinal ship's hull strength monitoring with optical fiber sensors. In: EMAR 2009, Zadar, 28–30. 09. 2009.
9. Cardis, P. *Inspection, Repair and Maintenance of Ship Structure*, Witherby Co, London, 2001.
10. Russo, M. Određivanje deplasmana prema gazu broda ianaliza dobivenih mrtvih težina radi ocjene točnosti po gazu utvrđene količine tereta. *Naše More* 1991, 38 (1–2), 33–38.
11. Ilyas, M., and Mouftah, H., *The Handbook of Optical Communication Networks*, CRC Press, Boca Raton, FL, 2003.
12. Murayama, H., Kageyama, K., Naruse, H., and Shimada, A. Distributed strain sensing from damaged composite materials based on shape variation of the Brillouin spectrum. *J. Intell. Mater. Syst. Struct.* 2004, 15, 17–25.
13. Hurni, A. et al. Fiber-optical sensing on-board communication satellites. In: Conference: International Conference on Space Optics, 2014.

Index